MAKING PCR

MAKING PCR

A Story of Biotechnology

PAUL RABINOW

THE UNIVERSITY OF CHICAGO PRESS | CHICAGO & LONDON

Paul Rabinow is professor of anthropology at the
University of California, Berkeley. His numerous
books include *French Modern: Norms and Forms of
the Social Environment* and, with Hubert Dreyfus,
*Michel Foucault: Beyond Structuralism and
Hermeneutics,* both published by the University of
Chicago Press.

The University of Chicago Press, Chicago 60637
The University of Chicago Press, Ltd., London
©1996 by The University of Chicago
All rights reserved. Published 1996
Printed in the United States of America
05 04 03 02 01 00 99 98 97 96   1 2 3 4 5
ISBN: 0-226-70146-8 (cloth)
        0-226-70147-6 (paper)

Library of Congress Cataloging-in-Publication Data

Rabinow, Paul
    Making PCR: a story of biotechnology / Paul
Rabinow.
        p.   cm.
    Includes bibliographical references and index.
    1. Polymerase chain reaction—History.
    I. Title.
QP606.D46R33   1966
574.87'3282—dc20   95-49103
                                        CIP

To Monsieur le Blanc

Contents

Introduction

*Permit me to take you once more to America, because there one can of-
ten observe such matters in their most massive and original shape.*

—Max Weber, "Science as a Vocation"

Making PCR is an ethnographic account of the invention of PCR,
the polymerase chain reaction (arguably the exemplary biotechnological invention to date), the milieu in which that invention took
place (Cetus Corporation during the 1980s), and the key actors
(scientists, technicians, and business people) who shaped the technology and the milieu and who were, in turn, shaped by them.
PCR has profoundly transformed the practices and potential of
molecular biology through vastly extending the capacity to identify and manipulate genetic material. It facilitates the identification
of precise segments of DNA and accurately reproduces millions of
copies of the given segment in a short period of time. It makes
abundant what was once scarce—the genetic material required for
experimentation. Not only is this material abundant, it is no longer
embedded in a living system. Cloning had made scarce genetic
material abundant, but its obligatory use of living organisms as the
medium of reproduction was also its limitation; PCR took a major
step away from that dependency. The step constituted a capital
advance in the efficiency and, more important, flexibility of genetic
intervention. PCR's versatility has been astounding; scientists have
produced new contexts and new uses with stunning regularity.
These uses have opened new avenues of research, which have in
turn proved amenable to new uses of PCR. In less than a decade,

| 1

PCR has become simultaneously a routine component of every molecular biology laboratory and a constantly improving tool whose growth potential has shown no signs of leveling off.

This book focuses on the emergence of biotechnology, circa 1980, as a distinctive configuration of scientific, technical, cultural, social, economic, political, and legal elements, each of which had its own separate trajectory over the preceding decades. It examines the "style of life" or form of "life regulation" fashioned by the young scientists who chose to work in this new industry rather than pursue promising careers in the university world. It includes the perspective of the company's business leaders, some of whom left the more secure fortresses of the multinational pharmaceutical world to engage in a riskier and potentially more rewarding enterprise (in terms of work, money, power, celebrity). In sum, it shows how a contingently assembled practice emerged, composed of distinctive *subjects,* the *site* in which they worked, and the *object* they invented.

While starting this project in 1990, I was often intrigued by, but skeptical of, the claims of miraculous knowledge made possible by new technologies supposedly ushering in a new era in the understanding of life and unrivaled prospects for the improvement of health. The weekly *New York Times* science pages rarely failed to announce that every new discovery or technical advance "could well lead to a cure for cancer or AIDS." These pronouncements seemed less like objective journalism than the prose written to attract venture capital. I took them as seriously as the counterclaims, appearing in other arenas, that the Human Genome Project was inevitably leading to eugenics. Although it was perfectly possible that both claims would prove to be true (or false), the proclamation of their inevitability seemed decidedly premature. The tone seemed misplaced. Regardless of what miracles or nightmares the future might hold, I shared the perspective that a new field of institutional arrangements and cultural practices was *emerging* in the biosciences.[1] It seemed eminently worthwhile to begin learning enough molecular biology to form my own understanding and to take responsibility for it. As an anthropologist, I am curious about the *form of life* in the making both within the labs and beyond (as tentative, divergent, and emergent as it is).

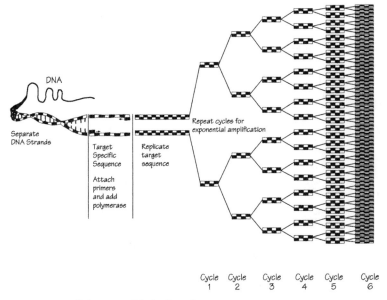

FIGURE 1. Polymerase Chain Reaction

WHAT IS PCR?

A few basics may well be helpful. What is a polymerase? A *polymerase* is a naturally occurring enzyme, a biological macromolecule that catalyzes the formation and repair of DNA (and RNA).[2] The accurate replication of all living matter depends on this activity—an activity scientists have learned to manipulate. In the 1980s, Kary Mullis at Cetus Corporation (the first recombinant DNA start-up company, founded in 1971) conceived of a way to start and stop a polymerase's action at specific points along a single strand of DNA. What is the *chain reaction?* Mullis also realized that by harnessing this component of molecular reproduction technology, the target DNA could be exponentially amplified. When Cetus scientists eventually succeeded in making the polymerase chain reaction perform as desired in a reliable fashion, they had an immensely powerful technique for providing essentially unlimited quantities of the precise genetic *material* molecular biologists and others required for their work (see figs. 1 and 2).

CYCLES	COPIES
1	2
2	4
3	8
4	16
5	32
6	64
7	128
8	256
9	512
10	1024
11	2048
12	4096
13	8192
14	16,384
15	32,768
16	65,536
17	131,072
18	262,144
19	524,288
20	1,048,576
21	2,097,152
22	4,194,304
23	8,388,608
24	16,777,216
25	33,554,432
26	67,108,864
27	134,217,728
28	268,435,456
29	536,870,912
30	1,073,741,824

FIGURE 2. Exponential Amplification

Though the simplest and most convenient way to define PCR is as a *technique,* such compartmentalizing eliminates the history of PCR's invention, thereby covering over the contingent manner of its emergence and the practices and subjects required to make it work. The next simplest answer is to name an individual as the inventor of the *concept.* The obvious candidate is Kary B. Mullis, who shared the 1993 Nobel Prize for chemistry for inventing PCR. However, as we shall see, this terrain is contested. Other scientists and technicians, including Henry Erlich, Norman Arnheim, Randall Saiki, Glen Horn, Corey Levenson, Steven Scharf, Fred Faloona, and Tom White, were instrumental in making PCR work.

A third argument holds that PCR did not exist until it was made to work in an *experimental system*. In this view, merely thinking of a concept is not sufficient; a scientific advance must include creating a way to show that the concept has successfully been put into practice.

Technique, Concept, Experimental System

When the prestigious journal *Science* named PCR and the polymerase it employs as its first "Molecule of the Year" in December 1989, the editor, Daniel Koshland Jr., provided a succinct explanation of PCR. Koshland and Ruth Levy Guyer describe PCR in the lead "Perspective" as follows:

> The starting material for PCR, the "target sequence," is a gene or segment of DNA. In a matter of hours, this target sequence can be amplified a millionfold. The complementary strands of a double-stranded molecule of DNA are separated by heating. Two small pieces of synthetic DNA, each complementing a specific sequence at one end of the target sequence, serve as primers. Each primer binds to its complementary sequence. Polymerases start at each primer and copy the sequence of that strand. Within a short time, exact replicas of the target sequence have been produced. In subsequent cycles, double-stranded molecules of both the original DNA and the copies are separated; primers bind again to complementary sequences and the polymerase replicates them. At the end of many cycles, the pool is greatly enriched in the small pieces of DNA that have the target sequences, and this amplified genetic information is then available for further analysis.[3]

After having described PCR entirely in terms of molecular biological technique, Koshland and Guyer conclude:

> The first PCR papers were published in 1985. Since that time PCR has grown into an increasingly powerful, versatile, and useful technique. The PCR "explosion" of 1989 can be seen as a result of a combination of improvements in and optimization of the methodology, introduction of new variations on the basic PCR theme, and growing awareness by scientists of what PCR has to offer. With

PCR, tiny bits of embedded, often hidden, genetic information can be amplified into large quantities of accessible,
identifiable, and analyzable material.[4]

For Koshland, PCR is a facilitating technology whose existence
can be dated to the first scientific papers in 1985. It took close to
four years for specialists to appreciate the technology's potential,
and longer still for a larger scientific community to begin practically exploring its power.

In 1985 *Science* published the first paper on PCR. In March 1986
Science turned down a paper submitted by Kary Mullis describing
PCR, sending him a standard letter of rejection: "The manuscript
did pass initial screening by our Board of Reviewing Editors, but
unfortunately the detailed reviews were not as enthusiastic as those
for other manuscripts considered at the same time. Thus, the paper
could not compete for our limited space."[5] In his "molecule of the
year" history, Koshland did not mention Mullis. The "molecule of
the year" had no author; Koshland said absolutely nothing about
who invented PCR. In the scientific literature there were as yet no
"geniuses," or even named inventors, associated with it.

In an account he gave to the Smithsonian Institution's Archive
of Biotechnology, Kary B. Mullis defined PCR not as a specific
technique, or bundle of techniques, but rather as a concept. For
Mullis, PCR came into existence at the moment he conceived of it.
For him, making the concept work was of secondary importance.
Mullis says:

> The thing that was the "Aha!" the "Eureka!" thing about
> PCR wasn't just putting those [things] together ... like
> saying you could denature, just could renature, you could
> extend, but it was to say you could do that three times and
> then this would happen.... *the remarkable part is that you
> will pull out a little piece of DNA from its context, and that's
> what you will get amplified. That was the thing that said, "My
> God, you could use this to isolate a fragment of DNA from a
> complex piece of DNA, from its context."* That was what I
> think of as the genius thing.... In a sense, I put together
> elements that were already there, but that's what inven-
> tors always do. You can't make up new elements, usually.
> The new element, if any, it was the combination, the way

they were used. . . . The fact that I would do it over and over again, and the fact that I would do it in just the way I did, that made it an invention. . . . the legal wording is "presents an unanticipated solution to a long-standing problem," that's an invention and that was clearly PCR.[6]

Mullis's thesis is partially plausible: he is correct that the specific techniques that composed PCR were not new per se. However, his general claim that technical elements are not invented is totally implausible. It is possible to date the technique for making oligo-nucleotides (short strings of bases of defined length and composition), the development of the electrophoretic gel on which DNA is made to migrate by an electrical current (the means used to separate out strands of different sizes), and the techniques used to transfer these strands to a membrane and detect them. What *was* original, powerful, and significant was the concept that combined—and reconfigured—these existing techniques.

Further, although Mullis claims that PCR was the solution to a long-standing problem, he never says what that problem was. A technician at Cetus, Stephen Scharf, is more perceptive when he says that the truly astonishing thing about PCR is precisely that it *wasn't* designed to solve a problem; once it existed, problems began to emerge to which it could be applied.[7] One of PCR's distinctive characteristics is unquestionably its extraordinary versatility. That versatility is more than its "applicability" to many different situations. PCR is a tool that has the power to create new situations for its use and new subjects to use it.

Almost everyone now agrees that Kary Mullis thought up the concept of PCR. Yet a group of former Cetus scientists and technicians maintains that only when an *experimental system* was developed did PCR become a scientific entity. In this view, PCR needed to be more than a series of disparate technical elements, and more than the synthesizing of these elements into a distinctly innovative concept. The concept needed to be practiced, producing results that met scientific standards. As Henry Erlich, a senior scientist at Cetus during PCR's development and now at Roche Molecular Systems, puts it: "Once PCR had been worked out, i.e. developed, only *then* was it useful."[8] During 1984 and 1985, two teams at Cetus were working on PCR: Mullis and Fred Faloona, a loyal assistant without a B.A.; and a team of senior scientists and the top technicians in their labs. The second team took half a year to produce credible and publishable experimental results and close to two years to produce a system of reagents and techniques that yielded experimental results specific and sensitive enough to begin to demonstrate the power and flexibility that Mullis's concept promised. Erlich and other Cetus scientists seem to agree with François Jacob's dictum: "In biology, any study begins with the choice of a 'system.' On this choice depends the experimenter's freedom to maneuver, the nature of the questions he is free to ask, and even, often the type of answer he can obtain."[9]

Invention

Who, then, invented PCR? As Norman Arnheim, a former Cetus scientist, replied when I asked him this question: "Conception, development and application are all scientific issues—invention is a question for patent lawyers."[10] At about the time *Science* named PCR the "Molecule of the Year," lawyers for the Du Pont Corporation were assembling a lawsuit against Cetus Corporation, challenging Cetus's patents on PCR. Du Pont claimed that there was nothing new in PCR and that all of its constituent elements had existed since the late 1960s, when they had been invented in the lab of Nobel Prize laureate H. Gobind Khorana.[11] The jury voted unanimously on more than fifty separate points to affirm Cetus's 1987 patents. Legally, the issue of who invented PCR had been settled. Did the jury make the correct decision? If, in fact, PCR is nothing but a set of *techniques,* then there is at least a plausible case to be made that the requisite skills existed to make the technique function before Mullis's 1983 conceptualization of PCR as a process of decontextualized, exponential amplification. Nevertheless, it is, at the very least, curious that for fifteen years after the supposed invention of PCR, none of the alleged inventors employed the technique, developed its variations, or patented it.[12] It is plausible to argue that they had the techniques but lacked the concept. In commonsense terms, it is obvious that PCR did *not* exist prior to Mullis's concept.

Invention is a question not only for patent lawyers but also for journalists, historians, Nobel Prize committees and those seeking such laurels, and anthropologists. PCR certainly might have been invented earlier by others: the necessary techniques and workable experimental systems existed. What was lacking was the concept. There was no inherent reason why the concept couldn't have been thought of during the 1970s, which leads us to speculate briefly on what factors might have been present to focus molecular biologists' and biochemists' attention elsewhere. One explanation is that techniques to manipulate DNA were still hierarchically dominated by concepts and systems in molecular biology and biochemistry. Khorana and his colleagues were constructing a gene; they wanted multiple copies of it. Cloning, which emerged in the early 1970s, provided the means to achieve that end—by harnessing known biological processes—yielding, if not *in vitro* exponential amplification, a sufficient number of *in vivo* amplified copies for the purposes at hand. Technology was serving biology. Although in hindsight it may appear that the scientists in Khorana's lab were close to PCR, the historical fact remains that cloning and other techniques solved their problem for them. Once techniques adequate to the task at hand became available to Khorana and his co-workers, they stopped exploring other possible means of amplifying DNA.

In an important sense, Mullis had no biological problem to solve (although others at Cetus did and it was within the context of work on the mutation of the beta-globin gene that Mullis conceived PCR). He was employed by Cetus Corporation to make oligonucleotides, a time-consuming and repetitive task. Henry Erlich captures this point by saying: "Khorana was asking a [scientific] question: 'Could I synthesize a gene?' To do so, he didn't choose any random 158 base pair DNA fragment. Mullis's job was making oligonucleotides. He would talk about making 'a 158-mer.'"[13] Genes were becoming manipulatable biochemical matter. Khorana was trying to harness a biological process (polymerization) as part of a larger project to make an artificial version of a biological unit, a gene. Mullis's decontextualization and exponential amplification was the *opposite* of Khorana's efforts at the mimicry of nature. Mullis conceived of a way to turn a biological process (polymerization) into a machine; nature served (bio)mechanics.

MANICHEAN COMMONPLACES

One of the hallmarks of biotechnology companies in the 1980s was precisely their attempts to produce a milieu that would facilitate exchanges between the university world and industry so as to minimize the cultural differences between them and to make a productive and profitable use of science. Such spaces allowed scientists to do their science (within limits) as they might have in the university world (whose own limits were converging with those in industry). The description Chandra Mukerji gives of oceanographers working on government projects rings largely true of Cetus's senior scientists:

> What reassures scientists the most when they face the power of the voice of science and their powerlessness in the public arena is the idea of their autonomy. Scientists are not, in the end, politicians, and they suffer political defeats better than loss of face among their peers. As long as they can conduct research with which they can advance their science (both science and their positions in it), they can feel potent. But the cost is that scientists cultivate an expertise that empowers someone else.[14]

In the case of Cetus and related companies, of course, it is not politicians but business people (and lawyers) with whom the tradeoffs were contracted.

It has become a commonplace that during the 1970s there was a blurring of the once supposedly distinct line between "applied" and "pure" research in the biosciences. Often this story is constructed as a corruption tale: how molecular biology became a handmaiden (to use a dated metaphor) to an industry relentlessly—and, by definition, unscrupulously—in search of profit. For example, one prominent historian of modern biology writes: "Prior to the emergence of recombinant DNA technology, the practice of molecular biology was guided largely by the traditional ethos of academic research and its commitment to the development of knowledge through an essentially cooperative and communal effort."[15] Anyone the least bit familiar with the history of science—the race to discover the molecular structure of DNA or the international battles to determine who deserved credit for identifying HIV, not to mention the career of Isaac Newton—could hardly describe it as an "essentially cooperative and communal effort." This view of contemporary bioscience—shared by many—is historically and anthropologically inadequate.

Scholars are providing rich accounts of the biosciences that underline the centrality of "applied" work and instrumental goals in modern science. For example, Evelyn Fox Keller analyzes the long lineage of attempts to control and manipulate "life," linking these efforts to the modern Western systems of gender and metaphysics.[16] Philip J. Pauly demonstrates the strong pragmatic and mechanist strand present in biological thinking and research throughout the twentieth century.[17] A growing body of research documents the history of industrial research in the chemical and pharmaceutical industries, which employed large numbers of scientists and technicians for well over a century.[18] Several recent books describe how medical research has been the object of repeated reform efforts directed by philanthropic foundations and later by government agencies.[19] Finally, it is well known how military research has taken an increasingly large share of governmental expenditures for science and technology during the cold war period.[20] Therefore, arguments claiming that the rise of the biotechnology industry constitutes an epochal breach of the traditional barrier between pure and applied research will have to be made a good deal more specific if they are to carry credence.

If one corruption tale recounts how capitalism corroded the pure coin of science, another genre of account seeks to show in a myriad of different ways that science was never what its practitioners and philosophers claimed it to be, that there had never been anything pure to corrupt. The contemporary field of the social studies of science was launched into its current trajectory by an attack, now largely taken for granted, on the view that there are distinctive scientific norms. The classic formulation of those norms is found in the work of the sociologist Robert Merton, who identified in the 1930s four interrelated, mutually reinforcing norms—universalism, communalism, disinterestedness, and organized skepticism. Merton argued that these norms guided the practice of empirical research and yielded "the extension of certified knowledge."[21] *Universalism* means that scientific truths are held to be impersonal, that is, they are the same everywhere and are independent of the scientist who makes a discovery and of the place of discovery. *Communalism* implies that science is a preeminently social activity building on previous efforts and shaping future ones. *Disinterestedness* refers to a commitment to truth above all other motives. *Organized skepticism* entails the social evaluation of truth claims through open debate, peer review, duplication of work, and the like. In Merton's picture, the main rewards for scientists' achievements are community recognition and prestige. The competition for prestige yields a pressure to be first but does not fundamentally falsify the knowledge produced. Quite the opposite: the ingenuity of the system for Merton is precisely that scientists, by acting in their own self-interest, collectively reinforce the public good.

Empirical studies of scientists at work revealed practices embodying conflicting values equally as important as the norms Merton stressed. In 1980 the sociologist Michael Mulkay published an article summing up the evidence critical of Merton's thesis. He observed:

> Common access to information is not an unrestricted ideal in science; it is balanced by rules in favor of "secrecy." Intellectual detachment is often said to be important by scientists; but no more so than strong commitment. Rational reflection is seen as essential; but so are irrationality and free-ranging imagination. The use of impersonal criteria of adequacy is often advocated; but the necessity of personal judgments is also frequently defended.[22]

Replication of scientific experiments, often held up as a central distinguishing mark, was shown to be less frequent and more partisan than expected.[23] Mulkay shows that terms like "adequacy," "consistency," and "replicability" are not Platonic norms that imitate preexistent forms for all eternity but, rather, are themselves situated constructs subject to negotiation, differential access, and other such social variables studied by sociologists of knowledge.

The explosion of work in the last two decades concentrating on laboratory practices rather than on the logical structure of theories, and on the negotiations and uncertainties of everyday science rather than on biographies of great men (and a few great women), has had an immense impact on how the sciences are understood.

The social study of science (and technology) is today a burgeoning multidisciplinary field. Beginning with the work of Thomas Kuhn and Paul Feyerabend and accelerating with the studies of Bruno Latour, Karin Knorr-Cetina, Evelyn Fox Keller, Donna Haraway, and many others, these diverse studies of the local practices of science have sought to bring the abstractions of Science, Reason, Truth, and Society down to earth. They have done so through empirical observation of laboratories and related research sites as well as through experiments in new textual forms. They have shown in diverse manners that these contexts—laboratories, writing genres, professional associations—contain many complexities that a functionalist sociology such as Merton's was incapable of incorporating. However, the contemporary sociological understanding of science has also produced, at least in some quarters, an inverse but parallel counterimage based on a conviction that, when scientists invoke norms such as "organized skepticism" or "disinterestedness," they are doing nothing more than donning a mask in order to conceal their base motives. If one corruption story leads its adherents to righteous nostalgia, at least some versions of this latter genre tend toward a self-satisfied cynicism. Ironically—and perversely—denying the very existence of normative elements because they are not completely and coherently instantiated, because they exist simultaneously in relationship with their opposites, amounts to denying that science is a social practice. For some in the social studies of science, unless science is pure it cannot be identified as a distinctive practice; for it to be pure it would have to exist outside of human activity. Q.E.D.

Although each component of Merton's picture of science has been subjected to historical, sociological, and philosophical reevaluation, it is fair to say that many scientists believe that these norms guide their practice. Hence, a major gap has developed today between scientists' self-representation and the representations of scientists by those who study them. While this discrepancy is of little consequence for practicing scientists (most will never have heard of its existence), it provides much of the subject matter and the authority for the social studies of science. Actually, the identification of the discrepancy is itself normative, a practice of producing knowledge. The fact that the self-image of the biosciences is not "true," i.e., fully adequate to its own norms, is precisely what makes it an appealing subject for anthropological study. I define practices as norms in context and in process. In that light one can identify contemporary inflections in the process of fashioning forms at least partially adequate to scientific norms understood as practices, with all the hesitations, conflicts, and failures attendant to such efforts.

Virtue and Veracity

In the epilogue to his consequential book, *A Social History of Truth*, Steven Shapin argues that the nostalgia of our modern story of the "world we have lost" blinds us to some enduring practices we have not actually lost. Shapin's book is an extended historical account of the emergence of a distinctive figure, the gentleman-experimentalist. This figure joined a new site of experimental practice and a culture of authority. In the seventeenth century English world of Robert Boyle and his air pump, where the emergent sites and actors of experimental science were taking on new norms and forms, scientific work and the gentleman's house were spatially adjacent. The credibility of experimental results produced in his laboratory was tightly linked to a gentleman's moral reputation. Knowledge was given authority through the earned trust of people who knew each other intimately. "Veracity was understood to be underwritten by *virtue*." The judgment of virtue, the according of trust, the authorizing of truth were part and parcel of personal networks, comings and goings, the conjoined spaces of everyday life. "Premodern society," Shapin puts it, "looked truth in the face."[24]

Today we are told by self-styled, hardheaded, suspicious moderns that such relations, such proximity, such evaluations, hence such virtues, have disappeared in the cold and anonymous world of macrobureaucracies and "big science." "[T]he modern place of knowledge," Shapin writes, "appears not as a gentleman's drawing room but as a great Panopticon of Truth."[25] Everyone is under surveillance. The foundation of peer review, after all, is anonymity. It is no longer virtue that guarantees knowledge but *expertise*.[26] We might add that such modern expertise is authorized by institutions, warranted by patent offices, legitimated by prize committees, and, on a different register, sanctioned by investors.

Shapin astutely points out that while this authority structure may well operate outside lab walls (and that claim, of course, is open to a certain caution), inside those walls the situation is not so dichotomous. Outsiders "tend happily to refer to vast numbers of practitioners called 'scientists,' insiders function within specialist groups of remarkably small size. . . . There is hardly any systematically collected information on the subject."[27] Within such core groups, face-to-face interaction and judgments of virtue or its lack continue to play a role. Radical distrust among practitioners may well carry consequences nearly as weighty as those it had for the early modern gentry; one's claims and concepts risk being prematurely, if not definitively, discounted.

For the anthropologist, Shapin's account rings partially true. There is no doubt that evaluations of character, and their consequences for trust and mistrust, figure centrally in science; they certainly shaped the early days of PCR. It is doubtful, however, that contemporary character evaluations contain more than residual echoes of the culture of Shapin's gentlemen-experimentalists. Though no longer at the center of a living system they once characterized, virtue and veracity today have been redefined and recombined with other mechanisms to produce different subjects, different objects, and a different milieu.

SCIENCE AS A VOCATION

In a celebrated address delivered on the day of the Bolshevik seizure of power in 1917, Max Weber analyzed the civilizational conditions undergirding "science as a vocation." He divided his essay into three sections: (1) The "external" or institutional conditions in comparative perspective. Weber contrasted the institutional career obligations imposed on German and American academics: the heavy teaching loads but better pay of the Americans, the American cultural consensus that the goal of knowledge was closely tied to utility, and the American professors' comparatively lower status (especially compared to football coaches, Weber remarks); (2) the power of conceptual clarification. The invention of the *concept* was, for Weber, one of the greatest achievements of the ancient world. Weber saw another qualitative threshold passed when, roughly two millennia later, rational experimentation as "a means of reliably controlling experience" (equally found in other civilizations) was transformed into a "principle of research"; (3) The "inner" rewards of a life of science, or science as a vocation (*beruf*, "calling"). Weber saw science entering into "a phase of specialization previously unknown ... that will forever remain the case." The fate (and the goal) of scientific activity was to chain itself to a course of progress that made individual achievements acceleratingly antiquated.

Given these conditions, Weber gravely summoned his audience to clarity about the life they were about to lead:

> Whoever lacks the capacity to put on blinders, so to speak, and to come up to the idea that the fate of his soul depends upon whether or not he makes the correct conjecture at this passage of this manuscript may as well stay away from science. He will never have what one may call the "personal experience" of science. Without this strange intoxication, ridiculed by every outsider; without this passion ... you have no calling for science and you should do something else.[28]

Add to this challenge Weber's observation that "Ideas occur to us when they please, not when it pleases us" and the question of what kind of person could endure, not to mention thrive in, such an environment, and it becomes clear that the scientific vocation retains its actuality, despite the melodramatic, neoromantic cast of Weber's prose. Whatever the material and career rewards a life of science provides, Weber's wonder at the historical, ethical, and existential particularity of the experience of experimental science remains pertinent.

In a more pragmatic tone, John Dewey, in his 1917 introduction to a book of collected essays entitled *Essays in Experimental Logic*, provided a credible credo that links experimentation and experience, human and natural science, practice and vocation. Dewey wrote:

> To place knowledge where it arises and operates in experience is to know that, as it arose because of the troubles of man, it is confirmed in reconstructing the conditions which occasioned those troubles. Genuine intellectual in-

tegrity is found in experimental knowing. Until this lesson is fully learned, it is not safe to dissociate knowledge
from experiment nor experiment from experience.[29]

Of course, the lesson is never "fully learned"; the "troubles" constantly reappear at each conjuncture. In my view, the task of the
human sciences is neither glorification nor unmasking, nor is it to
embody some phantom neutrality. The anthropologically pertinent point is the fashioning of the particularity of practices. A significant omission from the by now classic laboratory studies has
been the representation of science as a practice and a vocation—
by its practitioners. Such representations must be elicited, for there
are few other contexts in which they might be produced in a formal manner, i.e., given reflective form. Once elicited and given,
they must be framed in the light of larger forces at play (forces
whose impact many scientists are acutely aware of in the course of
their daily lives, even if they are rarely called upon to analyze
them). Thus, in addition to being an account of the invention of
the polymerase chain reaction, and the distinctive milieu from
which it emerged, this book is itself an experiment in posing the
problem of who has the authority—and responsibility—to represent experience and knowledge.

ONE

Toward Biotechnology

What would later be called the biotechnology industry emerged during the 1970s. The naming is itself significant because many of the projects the industry began with were old ones, some, such as fermentation processes, quite hoary. Whether projects such as beer making or producing antibiotics and vitamins should be considered precursors of the new biotechnology or holdovers from a previous time reflects the limits of epochal labels. Whether it was called "genetic engineering," "recombinant DNA," "cloning," or something else, the promise of a new era of efficiency and invention powered by scientific and technological advance provided the cachet and selling point for a series of diverse developments in science and commerce.

The main elements that contributed to the industry's initial appearance and the shape it has taken since, though familiar, are worth reviewing because they form the larger environment from which PCR emerged. These elements include: (a) the greatly enhanced technical ability to "recombine," "engineer," or simply "manipulate" DNA and other molecules; (b) a regulatory environment that encouraged the rapid application of research to applied problems, as well as changes in the patent laws directed at actively encouraging (almost forcing) the commercialization of inventions in both industrial and academic settings; and (c) the eventual dovetailing of governmentally funded research with venture capital looking for investments to form an expanded base for molecular biological research and development.

It was within this context of investments in scientific progress directly linked to new products and services in health that molecu-

lar biology came of age in the United States. The subtext of this
story is the advance of technology, the ability to manipulate DNA
under laboratory conditions. The importance of this subtext lies in
its exemplarity: the truths of molecular biology emerge from
model systems and the techniques used to create and study them.
Biotechnology's hallmark, it could be said, lies in its potential to
get away from nature, to construct artificial conditions in which
specific variables can be known in such a way that they can be
manipulated. This knowledge then forms the basis for remaking
nature according to our norms.[1]

A THRESHOLD CROSSED: 1979–82

In 1980 the Supreme Court of the United States ruled by a vote of
5 to 4 that new life forms fell under the jurisdiction of federal
patent law. General Electric microbiologist Ananda Chakrabarty
had developed a novel bacterial strain capable of digesting a component of oil slicks. Chakrabarty modified an existing bacterial
strain by introducing a new DNA plasmid (a circular form of
double-stranded DNA carrying a specific gene) into bacterial cells,
thereby giving the organism the capacity to break down crude-
oil components. In so doing, he produced a new bacterium with
markedly different characteristics from any found previously in
nature—one having the potential for significant utility. Chakrabarty having invented something "new, non-obvious and useful,"
the Court found it natural to protect his product with a patent.

A report from the U.S. Office of Technology Assessment con-
cisely underlined the dimension of the Court's decision that has
drawn the most public attention: "the question of whether or not
an invention embraces living matter is irrelevant to the issue of
patentability, as long as the invention is the result of human intervention."[2] The Court's decision thereby establishes a broadened
interpretation of the "product of nature" doctrine, which holds
that for a naturally occurring being or process to be patentable it
must contain a "substantially new form, quality, or property."
While plant forms had been patentable since 1930, a variety of
factors ranging from the organization of the seed industry to the
slowness of existing methods for breeding new plant varieties prior

to the advent of genetic engineering had contained the scope and impact of such patents until recently.[3]

Until the 1980s, patents had generally been granted only in applied domains. The language of the Constitution, which authorizes Congress to "promote the Progress of Science and useful Arts, by securing for limited Times to Authors and Inventors the exclusive Right to their respective Writings and Discoveries," had been interpreted to mean that patent law should promote the progress of "useful arts," that is, applied technologies.[4] Further, the Patent and Trademark Office had tended to restrict patents to operable inventions, not ideas; it had interpreted the Constitution to demand that an invention have a demonstrable and substantial "utility" in order to qualify for a patent. Finally, prior to the Chakrabarty case, it was generally held that living organisms and cells were "products of nature" and consequently not patentable. The requirement that patent protection be extended to the invention of "new forms" did not seem to apply to organisms (plants excepted). The patents on antibiotics had been awarded based on the isolation of these natural products in "pure form" rather than on the cells or organisms producing the antibiotics. "Nature" was public and available for all to use. The Chakrabarty decision validated a new dimension in the place "nature" would have in both the scientific and the cultural world.[5]

The Supreme Court's ringing proclamation that "Congress intended statutory subject matter to include anything under the sun that is made by man," coming as it did in the same year as the election of Ronald Reagan as president of the United States and the massive influx of venture capital into the biotechnology world, not only opened up "new frontiers" in the law but can be appropriately seen as an emblem of an emerging new constellation of knowledge and power.[6] Fredric Jameson's characterization of late capitalism as marked by its global reach as well as by the penetration of capital into nature on a transformatory scale never before possible is apposite.[7] Claims of a similar sort, albeit employing a different jargon and rhetoric, became standard fare in biotechnology companies' annual reports.

In 1980 Congress also passed the Patent and Trademark Amendment Act "to prompt efforts to develop a uniform patent policy that would encourage cooperative relationships between universities and industry, and ultimately take government-sponsored inventions off the shelf and into the marketplace."[8] At the time the government had some twenty-five different patent policies. This thicket of regulations tended to discourage exclusive licensing agreements, which in turn made industrial investment in product development less likely. The goal of the new policy was to encourage technological advance and a closer connection of university-based research with industry. Under the act's provisions, universities housing government-funded research were obliged to report any potentially patentable invention arising from that research. Failure to do so meant, under the so-called "march-in rights," that such rights automatically passed to the government.[9] The universities responded with enthusiasm. An Office of Technology Assessment report on *New Developments in Biotechnology: Ownership of Human Tissues and Cells* states that from 1980 to 1984, President Reagan's first term in office, patent applications from universities in relevant human biological domains rose 300 percent.

The Role of Government

Although debates about the social and ethical consequences of the biosciences often turn on the pivotal role of business, it is worth remembering that the initial major impetus for bringing applied and pure research in the biosciences into a closer, more productive relationship came from the U.S. government. Government research facilities and foundations, created or expanded after World War II, from the National Institutes of Health (NIH) to the National Science Foundation (NSF), established a vastly expanded presence, eventually taking precedence in shaping policy over the older philanthropic foundations that had played this role in key sectors in the years between the two world wars. Improving the health of the American people through medical research became a national policy objective. It also became an arena of major economic and bureaucratic growth.

For example, the so-called "war on cancer," proclaimed in 1971, directed substantial research funds toward achieving practical results in the fields of health. It also buttressed the dominant role of the federal government in directing biomedical research.[10] One has only to remember the largely private organization and funding of the campaign to discover the cause of and cure for polio in the 1950s to recognize the changes that took place in less than two decades.[11] By the end of the 1970s, the federal government "was pouring 11 percent of all federally funded research-and-development moneys into basic biomedical research, compared to 2 to 4 percent for most other developed countries."[12] However large such an investment in "health" may have been, it pales in comparison with the R&D budget allocated to the military, which remained constant at about 50 percent for most of the 1970s, rising to 60 percent in 1982 and 74 percent in 1987 (largely because of "Star Wars" and related high-tech weapons systems).[13] The purpose of such a vast outpouring of money was to achieve practical results, to orient research by defining an agenda.

The first major change in the funding of recombinant DNA research came from the federal government. In 1975 only two projects had been funded, at an approximate cost of $20,000; in 1976 the National Institutes of Health were sponsoring 123 projects, at an approximate cost of $15 million.[14] These were university-based projects. The motivation for increased spending was a combination of emergent and promising new technologies, such as cloning, and the spotlight focused on them by a growing national debate about their safety and broader social implications.

A Recombinant Configuration: 1974–79

A climacteric moment in the takeoff of the biotechnology industry was the extraordinarily swift and seemingly definitive resolution of the controversy over the safety of recombinant DNA. A whole range of technologies and research areas grouped under a single rubric became the focus of heated controversy that began in the early 1970s and peaked between 1975 and 1977. Because of an extremely rapid response by the leaders of the molecular biology world, by 1979 the debate had been all but silenced in the United States.[15] The importance of this outcome for stabilizing the institutional and commercial horizons of recombinant DNA work can hardly be overestimated. Had it taken significantly longer to put regulatory mechanisms in place; or had the types of experiments defined as low, medium, and high risk been different; or had safety controls requiring substantial capital investments been imposed for experiments defined as low risk; or had there been a sustained split in elite scientific opinion over the advisability of pursuing the technological and scientific lines of research and development in question; or had there been a major accident with recombinant material—however representative of the "true risks" it may or may not have been—the consequences for funding what was then called recombinant DNA technologies, in both the public and commercial domains, would surely have been dramatic. It is worth remembering that there was nothing ineluctable about the course events took.[16]

A preliminary debate on safety surfaced almost simultaneously with Paul Berg's successful recombinant DNA work in the early 1970s. This debate, originating in Berg's own laboratory, turned on determining what constituted laboratory safety. It included concern over the possible threat to environmental or public health by waste disposal or escape of recombinant organisms and materials. It expanded to more political and philosophical questions: what were the possible ecological and evolutionary implications of transferring DNA from one species to another? An initial voluntary moratorium suggested by Berg himself was followed in subsequent years by a series of high-level meetings attended by the leading players in the field. The emblem of these meetings, debates, and conjectures has now become the Asilomar Conference, held in February 1975. These meetings and consultations culminated in controls for recombinant DNA work in the U.S., Britain, and other countries. They represent a unique event in the regulation of technological applications. Elite scientists evaluated a new technology and developed regulatory guidelines that were subsequently adopted by the government. The regulations proved to be a very successful preemptive strategy that warded off further regulation by outsiders (to the scientific community or its elite informal groups). The molecular biology community's quick action assuaged many doubts (particularly in Congress), although it clearly did not provide definitive answers to a whole series of issues concerning either general risk analysis or broader ethical implications.

By 1977 and 1978, the argument prevailed that recombinant DNA was a supervised and safe technology as well as a potentially invaluable research tool with a huge potential for both profit and the improvement of health. As experience accumulated, work accelerated, and money was invested, the recombinant DNA guidelines were relaxed twice in 1980, again in 1981, and again in 1983.[17] By this point, "The simultaneous growth of small biotechnology start-ups financed by venture capital and the increased interest of multinational corporations created a backdrop of intense entrepreneurial activity for the congressional hearings and NIH guideline revision meetings."[18] Broader host-vector guidelines as well as a distinction between large-scale and small-scale experiments were introduced. Secret sessions of the Recombinant Advisory Committee (R.A.C.) were allowed in order to afford protection for proprietary interests. A corner had been turned. The safety issue had been contained. Government regulators, Congress, business, and a significant sector of the scientific community were satisfied.

Strategy: Patent and Publish

The early 1980s was a time of fierce competition during which the key arena for patenting and publishing was newly cloned genes. The status of the law in these areas was unsettled. Traditionally, in the academic system, publication and priority established scientific reputation. For business, the symbolic rewards for first publication were clearly secondary to the potential commercial advantages a patent offered. The management of biotech firms saw that publication could effectively prevent others from patenting a discovery. This realization led to the strategy of filing a U.S. patent application on a cloned gene and then publishing as rapidly as possible. Filing a patent application for an aspect of the work and subsequently publishing would serve as a means of establishing "prior art" and consequently barring others from obtaining a patent— especially outside the United States.

This strategy differed from older industrial strategies for success in cases where a slight variation in the structure or synthesis of a particular chemical might well be enough to get around a basic patent. Consequently, chemical structures were held secret for years until patents revealing that information were issued. The field of genetic research was so competitive that if one didn't publish findings quickly (often within a matter of weeks or months), there was a constant prospect that some other group would. Since the gene's cloning would be published anyway, the only effect of not publishing was to demoralize the company's scientists and to create the appearance of being second-rate.

A second aspect of this patent-and-publish policy was its use as a means of lessening the differences between the university and industrial milieux, a formerly large gap in the biosciences, closed or narrowed as a result of action on both sides. Many university scientists, led by the elite, responded eagerly to the new patent policies. Since the information was going to be made public rapidly in any case, it was beneficial to industry for its scientists to get credit in the larger world for their discoveries as a means of attracting and keeping high-level achievers. Further, such a policy facilitated work with university scientists by allaying the most commonly stated fear of those scientists that secrecy would be imposed for purposes of commercial gain.

Finally, the teams of bioscientists in companies like Cetus or Genentech often had a competitive advantage over their university colleagues because they were working in large teams with considerable resources of space, equipment, and staff flexibly available for timely use on specific projects. The scientific and commercial strategies concerning the flow of information were reflected in the design of biotech facilities. One of the architects hired by Biogen to design its new facilities observed: "To reinforce management's philosophy that the company is a synthesis of science and business . . . we designed several informal meeting areas—for coffee, reading, etc.—where staff could gather and exchange ideas. Meeting rooms are glass enclosed to convey a sense of open communication." [19] Of course, "a good physical environment is certainly to be desired, but it is not as important as a good informational environment. Scientists must be immersed in a constant flow of information and must be active participants in this process." [20] Internal seminars were attended by attorneys alert for patentable ideas; all outside visitors attending the seminars were bound by confidentiality agreements.

The nascent biotechnology world was the site of several important boundary renegotiations. The cloning and characterization of a new gene in an academic setting was considered to be basic research. In a company, on the other hand, it might well be considered to be both basic (since no academic lab had yet cloned it) and applied research, simply because of funding, project definitions, and potential therapeutic utility. The interdisciplinarity of the molecular biology–based biosciences was strategically spearheaded in the start-up companies. Martin Kenney, an early, and highly perceptive, analyst of biotechnology, observes that several large multinational corporations followed the lead of these startups during the early 1980s. More interesting, he speculates that these new modes of organizing research and development indirectly (and sometimes directly) provided pressure for some of the academic reorganization of the biosciences during the 1980s.[21] Finally, Kenney points to the key role of senior scientists in startup companies as the vital communication link between corporate management (and the scientific advisory board) and the bench scientists and technicians carrying out the projects the senior scientists had chosen and designed.

Symbolic and Material Capital
In the wake of the continuing technological advances as well as the stabilization of regulatory issues, it is fair to say that by the end of the 1970s a threshold was being crossed, a page was being turned. "The cumulative equity investment in all types of biotechnology companies rose from fifty million [dollars] to over eight hundred million between 1978 and 1981."[22] The various granting agencies of the U.S. government, backed by a supportive Congress and advised by the elite of the biosciences themselves, who were increasingly committed to this direction of research and development, raised funds for recombinant DNA work by 34 percent a year from 1978 to 1982. Again, these moneys went primarily to university research. Capital also began to flow from other major sources. These included a variety of multinational corporations, either giant pharmaceutical companies adopting a primarily defensive and hedging strategy lest they be left out of a potentially burgeoning new arena, or large companies that sought to diversify their investment portfolios and that had in mind specific projects for which they sought the expertise of small start-up companies. The central players in the birth of the biotechnology industry were venture capitalists and start-up companies. As their name implies, venture capitalists invested money in unproven arenas in search of high return. Most recently, new technologies had meant the computer and integrated circuit industries initially clustered in

the U.S. around the San Francisco and Boston areas, with their large pools of scientifically skilled laborers and rich and diverse academic environments. These same two areas were also the birthplaces of most of the start-up biotech companies.

The need for guarantees that investment money was being wisely invested was acute in this fledgling industry, given that the appearance of moneymaking products, even in the most enthusiastic of business plans, would be years away. The gold standard of legitimacy for investors was scientific prestige. Martin Kenney points to the novelty of the situation: "In less than a decade . . . a new industry and a new labor force ha[d] been created, and at the center of this maelstrom of activity were 'pure' scientists— molecular biologists. . . . The pervasive role of professors in managing and directing the start-ups is unique in the annals of business history."[23] Again, there was nothing new about scientists with advanced training and degrees being affiliated with industry; what was new was that this labor force was spearheaded by professors or by those whose career patterns and life-styles would have led them into the academy even a decade earlier.

In addition to traditional motives of moneymaking and the like, newer forces motivating prestigious scientists to embrace these companies (in different capacities) are to be found in the changing conditions of university research. Kenney points out that with the growth of the federal funding agencies, one of the basic requirements for a successful career in the biomedical sciences was the ability to secure grants and the associated skills in bureaucratic politics and thinking that grantsmanship entailed. Some studies estimate that 30 to 40 percent of a researcher's time could be devoted to the grant application process.[24] Leading scientists had become accustomed to spending large amounts of their time away from the lab benches, negotiating the increasingly large sums required to staff and run an up-to-date laboratory, consolidating the network of prestige and power upon which reputations and further funding depended, and spotting or promoting the "hot" areas toward which federal funding would be directed and around which scientific prestige accrued. Given this context, it is no wonder that the prospect of spending less time in pursuit of grants combined with the promise of substantial monetary rewards (either) immediately or in the future, from stock options) would prove attractive to many scientists.

Nobel Prize winner Arthur Kornberg sums up the situation well.

> Able scientists are interested in industry. Some are discouraged by the atmosphere often encountered in university departments: the emphasis on entrepreneurial skills of grantsmanship, the inevitable clashes with university bureaucracy, the obligation to serve on committees, the burden of heavy teaching loads, and the pressure to choose a safe, fashionable research program that will produce publications for the next grant application and academic promotion. In the face of these problems, one might see an industrial setting as offering several advantages: excellent resources, research objectives in interesting areas of science, fewer distractions, and a team spirit united for achievement.[25]

The cornerstone of this change—the redrawing of the lines between industry and academy, the convertibility of symbolic and material capital—was its sanctioning by the elite of the biosciences, the growing ranks of Nobel Prize winners who early on enthusiastically provided legitimacy for the redefinition of science and industry, even if this was not their primary motivation.

Nobel Rewards

An early distinctive feature of start-up biotech companies was their recruitment of well-known scientists to serve as advisors. The motivations for Nobel Prize winners to join clustered around the venerable triad: money, power, progress. Scientific advisors were remunerated with direct payments (on the order of ten to thirty thousand dollars a year for several meetings and telephone availability). They were also usually given large grants of shares of stock of nominal value prior to a company's first public offering or later stock options that might or might not pay off but had the potential to be extremely lucrative. Observers agree that in the eyes of these small groups of Nobelists in the biosciences—many from relatively modest backgrounds—money had no stigma whatsoever attached to it.[26] The duties of a scientific advisory board (SAB) were not onerous, often amounting to occasional meetings with old acquaintances, giving advice on projects over the phone, or "being available" for consultation. Although start-up companies with their venture capital offered their advisors the opportunity to make their ideas practical, contempt for the business management was often expressed by members of scientific advisory boards. Higher causes were being served. The secure tone of Arthur Kornberg's characterization of the academy-industry relationship in the following quotation is typical:

> We recognize that science and technology are interdependent and often inextricably linked. We know that advances in science depend on techniques; the availability of techniques in turn depends on innovative and aggressive commercial development. When sophisticated instrumentation and fine biochemicals become commercially available and affordable, research is extended a thousandfold.[27]

Kornberg expresses no doubts about hierarchy and rewards: "The scientists, departments, and universities that provided the ideas and reagents, the techniques and machines, and the very practitioners of genetic chemistry and immunology are reluctant to be excluded from financial rewards they very much need.[28] He mentions no ethical compromises or other such dangers arising from these arrangements.

Second, there was power and influence: "These guys were quite arrogant, absolutely sure they knew better, and more than willing to give advice on how to proceed."[29] There is a link here to another advantage available to members of SABs. The biotech companies were outlets where distinguished scientists could place their doctoral and postdoctoral students. At the end of the 1970s in a shrinking job market, this was particularly attractive, especially given the fierce competition for jobs in the San Francisco Bay Area and Boston. These younger scientists were then both advocates for, and workers on, the projects within the company that the advisors proposed. Political influence within the company itself for new projects could be affected by whether the younger postdocs had an ally on the SAB.

Finally, the prospect of having their discoveries count in the larger world, especially the world of health, was appealing to this largely male cohort, who had achieved everything else the scientific life could provide. There was apparently some degree of discomfort that ideas in university settings rarely led directly to health-oriented results. Throughout their careers, senior investigators had filled in the section on their grant applications (especially to the NIH) concerning the practical utility of the proposed research for benefiting health. Hence, it was logical for these SAB members to seek a commercial outlet for the research, which neither the university nor the government nor the big companies would provide. It was common during this period for these distinguished scientists to attempt to direct these projects as well. Such arrangements included grants from the company to support research in the advisors' university labs on the very projects they recommended.[30]

For the elite, it seemed to be the best of both worlds: there was the ongoing work of research and a tenured position at the university, there was the possibility of receiving substantial additional funding for one's own research from the companies one was advising, there was the prospect of doing good, there was an outlet for one's students, and there was money. These arrangements were beneficial to both the company and the university scientists. The company could raise funds by marketing the reputations and credibility of its advisors. The professors could not only obtain funds for their labs but also feel the pride, excitement, and gratification of knowing that their important work was being recognized by additional funding, albeit on their own recommendation.

CETUS CORPORATION

Cetus Corporation was founded in 1971, initially in the manufacturing area of Berkeley near the San Francisco Bay. Ronald Cape, a biochemist, and Peter Farley, a physician, both holders of M.B.A. degrees (unusual at that time), joined Donald Glaser, who had won a Nobel Prize in physics in 1960 for his invention of the bubble chamber (and who later became a molecular biologist), in founding the company. Initially, Cetus undertook a range of projects such as improving the yield of vitamin and antibiotic fermentations. Later, it began work on therapeutic human proteins and animal vaccines and produced genetically engineered yeast for alcohol production.

Cetus's business strategy consisted in making contractual arrangements or joint ventures with larger companies seeking to augment or complement the bioindustrial expertise available within their own firms. The vitamin B_{12} project typifies Cetus's operations in the early period. Although its goals were traditional pharmaceutical ones, Cetus's approach to achieving them was innovative and successful. A French pharmaceutical company, Roussel-Uclaf, produced vitamin B_{12} by fermentation of bacteria in twenty-thousand-liter tanks. The process was difficult because it began anaerobically (without oxygen) and then switched to aerobic conditions. The yield of B_{12} had been steady for years. Roussel-Uclaf's aim was to find new bacterial strains that were more efficient and more productive in making B_{12}. The traditional approach involved mutating many strains of bacteria each day, placing them in petri dishes by single colonies, and then screening hundreds of strains in flasks and testing those that proved to be slightly more productive. Such an approach was not very efficient and had not been substantially altered for decades, but it did produce a slow and steady increase in productivity. Roussel contracted with Cetus to explore new strategies. One approach Cetus developed combined classical molecular genetics and synthetic chemistry. Cetus chose to make mutants in bacteria that no longer made the vitamin; those nonproducing strains were remutated in order to make them productive. The method sought to out-compete, as it were, the random mutation approach. The goal was simply to increase production by reinvigorating the bacterial pool. The basic strategy, however, was the same as traditional methods. Cetus scientists attempted, with some success, to add more innovative recombinant approaches. Although the science contained some innovative aspects, the business side did not; Cetus scientists worked under strict confidentiality agreements, and publication was discouraged.

In addition to such pharmaceutical ventures, Cetus entered into joint ventures with affiliates of major oil companies to produce bioengineered products such as xanthan gum for oil recovery. A major influx of money resulted from a joint project begun in 1979 with Standard Oil of California to develop a new process to produce fructose from low-cost, corn-derived glucose. Fructose is a naturally occurring form of sugar found in honey and fruits and vegetables; it is sweeter than sugar but more expensive to produce in quantity than cane or beet sugar. The $11-billion market in sweeteners was attractive. During the late 1970s Cetus also was involved in some innovative work using *Bacillus subtilis* as a model system in an attempt to improve on the long-standing, standard model bacterial system, *E. coli*.[31] Although the *E. coli* was well understood, it presented specific problems for products intended for use in clinical settings, and Cetus hoped that *Bacillus subtilis* would provide a safer testing and production system. Initial results were promising, and the NIH granted Cetus permission to produce interferon using *B. subtilis,* but ultimately the impact of the new system proved disappointing.

Cetus Goes Recombinant

Only later, in the late 1970s, would Cetus become a recombinant DNA company. Even though major technological advances were taking place during the 1970s, it took time for them to become commercially viable. For example, only by the end of the decade had the parameters for the commercial application of cloning techniques been much improved and components such as reagents, vectors, host strains, and restriction enzymes become widely available. In 1979 Cetus signed a contract with Norden Laboratories, a subsidiary of Smith-Kline, to produce a vaccine to prevent colibacillosis, a diarrheal disease affecting newborn calves and piglets. This vaccine work encapsulates the elements that would define the biotechnology industry in the subsequent decade. The idea for the project came from a professor, Stanley Falkow of the University of Washington. Falkow had cloned two genes of an organism that produced a toxin that affected piglets. Instead of making a vaccine by traditional methods, from a killed pathogenic strain or weakened strain used to induce an immune response providing protection from the virus, he proposed a new method: cloning an inactive form of the toxin. Cetus succeeded in developing a vaccine strain, the first recombinant vaccine to reach the U.S. market.

During the late 1970s there was a major internal conflict among

Cetus scientists between the "traditionalists" and the "cloners," to use the terms employed by the actors themselves. This battle for power in the company became personified as a struggle among managers advocating contrasting strategies. One camp saw the future of the company and the industry in recombinant DNA work; another group felt recombinant DNA was at best a passing trend and that bacterial strain improvement programs and enzymology would be most valuable commercially; a third faction thought biotechnology would be useful only as a more efficient method for making chemicals. Major investors, potential joint-venture partners, and the investment bankers who handled Cetus's public offering were mainly excited by the potential of genetic engineering. By the turn of the decade, cloning had won. Once the technological agenda was more or less settled, there was a good deal of internal contention about what shape the research agenda should take.[32]

During the late 1970s and early 1980s at Cetus and other companies such as Biogen and Genentech, the primary research effort was devoted to genes that were to be cloned and expressed for possible use as therapeutic agents. These generally fell into the category of genes that coded for the production of proteins that were replacement factors for diseases caused by protein deficiency and that were therefore already in therapeutic use, such as human insulin, growth hormone, and Factor VIII (a blood-clotting factor). The utility and marketability of these compounds were established, but some were in short supply and some, produced from nonhuman biological materials, had undesirable side effects in humans. For example, cow and pig insulin had been used since the early 1900s to treat diabetes. The idea was to replace animal insulin with human insulin. Two methods were employed to secure supplies of human insulin: either cloning the gene for human insulin or synthesizing the entire gene chemically.[33]

The second category of genes for potential therapeutics consisted of those that coded for proteins, or mixtures of proteins, that had been partially purified from blood, tissue, or cells, and for which a plausible clinical effect could be proposed based on their *in vitro* (in a culture medium outside a living organism) or *in vivo* activity. Examples included the beta and alpha interferon(s), proteins that had antiviral activity and inhibited the ability of some transformed cell lines to grow *in vitro*. The latter effect was the basis for hypothesizing an anticancer activity of the interferons. Medical researchers eagerly sought to collaborate with any lab that had even minute amounts of protein, because purified proteins were in rare supply.

The third class of cloned genes were for "factors," so called because they were thought to be involved in the growth and differentiation of other cells. This group of proteins was initially somewhat disreputable from a biochemical perspective, since by definition they were characterized by a vague, often indirect biological activity on a cultured cell line, and the activity was not even clearly due to the effect of a single biochemical entity. The best characterized, and therefore first cloned, of this group was interleukin 2 (IL-2), which loomed large as a potential anticancer agent in shaping Cetus's research agenda.[34]

From the perspective of the business, genes in the first group (replacement factors) were better candidates because their medical importance and clinical efficacy were already known. Except for the blood-clotting Factor VIII, however, they were considered to be unexciting and to have a potential for only a restricted market. Genes in the second category (antiviral proteins) carried a higher risk, since their clinical value was unproved. Yet because there were no approved antiviral compounds at the time, the promise of an era of antivirals that would parallel the era of antibiotics in their therapeutic and commercial importance was a strong impetus to committing the scientific resources to clone the interferons. This promise was also the basis for the flow of the increasingly large amounts of money required to carry out the research. Genes in the third group—the "factors"—were by far the most risky for the companies. There was a scramble to locate academic collaborators who might have cell lines that could be used to isolate protein or messenger RNA directly or to assay their activity. Animal models were proposed to test for potential effects that would be desirable in human clinical trials. Of course, these efforts were subject to the criticism that the models might have no predictive value for human disease, but it was easy to lose sight of this reservation as momentum built.

The first annual report of Cetus Corporation in 1981 was predictably upbeat. It boasted of the wide scope of the company's activities, insisting that Cetus was "not merely a recombinant DNA company" but continued to take advantage of its "broad base of biological skills . . . microbiology, biochemistry, organic chemistry, monoclonal antibody technology, fermentation and process engineering, instrumentation design, and automated bioscreening."[35] At Cetus, "microorganisms are 'factories' producing useful products." The biotechnology industry represents, the report continues, "an opportunity for the creation of value and profit with few precedents in modern history."[36] Although the rhetoric was compelling and attracted money, the achievement of these 1981 bacterial "microfactories" did not measure up to their trumpeted potential. Still, between the hype necessary to raise money and the work required to make advances, carried out under conditions of intense competition, few actors afforded themselves the luxury of doubt. Cetus's pride in the breadth of its activities conceals what it would soon be forced to admit was also its major weakness—the lack of a distinctive and focused strategy for the future.

Company Milieu

Several characteristics of the Cetus milieu distinguished it from the older pharmaceutical corporate environment as well as from the academic world of molecular biology or biochemistry in the late 1970s. Cetus's organizational structure was less hierarchical and more interdisciplinary than that found in either corporate pharmaceutical or academic institutions. In a very short time younger scientists could take over major control of projects; there was neither the extended postdoc and tenure probationary period nor time-consuming academic activities such as committees, teaching, and advising to divert them from full-time research. Further, entrepreneurial firms generally avoided the staid, bureaucratically saturated procedures of older, more established industry. Interdisciplinarity was not merely encouraged as an abstract good, it was tacitly mandated in the sense that work was problem-driven and rewards were assigned for solutions. In contrast to the academy, where a scientist had to define a career in one particular field, and where the criteria for establishing a reputation were clear-cut, Cetus and related companies maintained during this period a relatively amorphous evaluation system, drawing on norms from both industry and the academy, which included such disparate criteria as discoveries and "cooperative spirit."

Standards for setting up timetables for cloning an unknown gene or for product development were often hypothetical because assays for biological function, expression methods for active proteins, and purification criteria for recombinant proteins were largely unknown territory. Goals were often a function of how much and what kind of funding was available. But since venture capital funding came to be increasingly a function of "promise," the demonstration of new discoveries and "progress"—however remote from commercialization—was often sufficient to keep money and rewards flowing. The distribution of credit and reward among members of an interdisciplinary team assembled to address a problem had to be negotiated, but as there was neither a fixed number of positions nor a stabilized monetary reward system, there was a great deal of flexibility in this realm as well.

INTERVIEW: DAVID GELFAND

In 1976 Ron Cape and Peter Farley, Cetus's top executives, invited David H. Gelfand, then a postdoctorial fellow at prestigious University of California San Francisco (UCSF), to head a proposed recombinant DNA division at Cetus. Gelfand responded with what he considered to be a utopian plan. To his surprise, Cetus management accepted his proposal. After implementing it, though, Gelfand soon tired of administration, finding that its time demands conflicted too strongly with his scientific interests and didn't suit his temperament. He returned to the bench. One might say that Gelfand represents a recombinant-era version of an older industrial scientist: pleased not to have to attend to heavy administrative duties, university publishing expectations, and grant deadlines.

PAUL RABINOW   David, could you start with your family background?

DAVID GELFAND   I guess left of center. I grew up in White Plains, a suburb just north of New York City. My father was an accountant. And a pro bono attorney for American Civil Liberties Union and Lawyers Guild and Lawyers Constitutional Defense Committee. And so I also had a politically active upbringing in terms of constitutional law and civil rights. My father was an Abraham Lincoln Brigade member. In high school, the most significant political activity was when we tried to get the principal to allow us to have an advanced biology class. I became independently interested in science, but was also very interested in American history.

PR   Why did you choose to go to Brandeis as an undergraduate?

DHG   Two reasons: of the places I had visited and applied to, it was one of the few that encouraged undergraduates to work in the laboratories. And because I'd done lots of experiments and taken summer courses at NYU and the University of Michigan, I was interested in doing research. I was also interested in politics. I wasn't sure whether I was going to be a scientist or a lawyer. In any event, I was involved at Brandeis in the Northern Student Movement, also in SNCC and also with SDS [Students for a Democratic Society] in 1962 or '63. I don't remember how I first became involved with the Student Nonviolent Coordinating Committee, but I remember one winter going into a demonstration outside of Albany, Georgia, and had come to know people involved in SNCC.

PR   Did being Jewish play an important part in your choice?

DHG   No, not at all.

PR   During this period Kennedy had just been assassinated and the beginnings of the escalation of Vietnam were taking place, Martin Luther King Jr. was moving onto center stage, you're deeply engaged in political activity. What were you doing in biology?

DHG   Well, they're separate. I'm not sure we thought about it at the time, although maybe we did. A conscious political decision to get northerners involved in Mississippi in the summer of '64, expecting that the violence that was being perpetrated on the SNCC staff people in '62 and '63 would also be visited on the northern volunteers, and that would carry with it the press and national attention and, hopefully, the army to take over the state of Mississippi. I don't know why we thought the army would help. Even though I had spent time with SNCC in the South, I really had not been prepared for the intensity of the violence. There were several incidents. One was at the courthouse where I had been in a proceeding because several of the other staff volunteers from the Laurel office had been attacked at a lunch counter. I'd been there to offer moral support and testimony if necessary. While there, I had observed the local head of the [Ku Klux] Klan attacking another volunteer as he was bringing someone to register to vote in the courthouse and had filed an assault and battery complaint against the head of the local Klan. A week or ten days later, we happened to be at a lake at a farm of a black family. This fellow and another dozen guys came out of the woods with chains and clubs, came up to me and said he was going to kill me. After being clubbed and shot at, I managed to get back to the farmhouse. We phoned the FBI to come rescue us. The local FBI and northern FBI agents, who were by this time in Mississippi, said they were not a protective agency and they could only investigate civil rights violations. I remember saying, "Fine, could you please interpose your body between the bullets and us so you could investigate the bullets crashing into the wall." They never came. Eventually I was taken to the hospital in a hearse from the black community because the white ambulance wouldn't come to pick me up. Since I knew who had attacked me, I had lodged a complaint for assault and battery, with intent to commit murder. The following fall I was scheduled for a grand jury hearing, and I went back to Laurel, Mississippi. An attorney from the New York Lawyers Constitutional Defense Committee came with me. We had a lot of difficulty getting our own depositions and statements made to the FBI from August, that I had made to use in this hearing. The hearing then got moved and reassigned to a backwater county seat rather than the city county seat, where our congressman had assured us we would have protection. After the grand jury hearing, a couple of cars tried to overtake us and run us off the road. And at that point, I decided that I was no longer going to pursue an American history/political science major. I wasn't averse to going to law school. I assumed that anyone who was interested in constitutional law, civil rights, and civil liberties law was going to be shot at and perhaps killed. Indeed, the lawyers who had helped us had bricks thrown through their hotel windows. While it's fine for someone who's a sophomore in college, it's not the way to grow up and live your life. I thought someday I'd get married and someday I'd have kids. I had blinders on.

In my junior year, I had begun to pester a cell biologist by the name of Gordon Sato to work in his lab. Finally after two years, he finally caved in and said yes. He was more of a mentor than my undergraduate advisor in the biology department. When it came time to decide what to do for graduate school, he advised me to choose UC San Diego. I had applied to two places: the University of Miami, Coral Gables, and the University of California, San Diego, because I was interested in scuba diving.

PR   Would it be fair to say at this point that, on the one hand, the Mississippi experience pushed you both towards a kind of privatization and, on the other hand, the consolations and joys of the republic of science? Science was going to be another kind of community, another kind of way of leading a life, that you still had faith in?

DHG   Yes, although I was still involved with SDS. On to San Diego! Gordon Sato told me I should go to San Diego—so I went to San Diego. One of his former postdocs was on the faculty there and arranged a summer job for me. It turned out that various people from SNCC and several folk musicians from the South lived in Del Mar. Since I'd played guitar in some of the clubs in Boston and Cambridge, I continued to play in San Diego. We started an off-base coffee shop in Oceanside, a folk music club, a rock place, geared to the cadets, the seventeen- and eighteen-year-old kids who had enlisted in the marines. We hoped to try to get them to think about what they were doing, and the club was called The Sniper. It drew the ire of the local marine base, but it was off base, and we just had folk music and a lot of Phil Ochs and Len Chandler folk music.

I was doing bacteriophage molecular biology. Classical. Heavy-duty molecular biology. I had done rotation projects in various labs, one with David Baltimore at the Salk Institute. We spent a lot of time discussing politics and an awful lot of time studying polio virus. My thesis was very intense because my thesis advisor, Masaki Hayashi, believed that rank hath its privilege:

graduate students must work harder than the professor. He worked from 11:00 a.m. to 4:00 a.m., seven days a week. He expected graduate students to finish in three and one-half to four years. No five-year theses. There were long hours all the time. There was always the competition. I liked him a lot. His lab was a wonderful lab to be a graduate student. You had to learn to do everything yourself. It was not a very good lab for a postdoc because he didn't go to many meetings. He was not a part of the network.

PR   Was there any talk of industry at this point?

DHG   *Absolutely Not!* Not *even* a remote chance. I finished up, staying on as a postdoc in the same lab because my wife had enrolled in graduate school in the sociology department at UC San Diego for a Ph.D. program. I stayed for two years as a postdoc while she was going to graduate school. The draft was affecting us also. As I recall, the draft law changed in July of '66. I had had a 2-S deferment that expired June 30, 1967. My draft board had failed to classify me, even though I was a graduate student. They subsequently sent me a 1-A classification, because you had to ask for a 2-S. The first thing I did was go to Masaki and arrange that I could complete my Ph.D. at the Pasteur Institute. I was *not* going to Vietnam! By that time, it must have been spring of '68, my wife was pregnant. I was able to obtain a 3-A deferment. Then I moved to UC San Francisco in January of '72 to work with Gordon Tomkins. I had met Gordon Tomkins at the Cold Spring Harbor symposium in 1970 and thought he was just wonderful. We had talked over the next year and a half, and it worked out that I would come to his lab. And three and one-half years after coming to Gordon's lab, he died tragically.

In April of '76 I had a phone call from Ron Cape, saying that he was president of Cetus, a company in Berkeley, and heard that my future at UC was indefinite, that I was looking for a job. I said, "Well, Ron Cape, president of Cetus, my future at UC is not indefinite: I can stay here for the next five years. I'm not looking for a job. Under no circumstances would I consider leaving before nine or ten months from now. I thought that Cetus was doing something biological in the East Bay. Occasionally I have lunch, so if you come over I'll be glad to have lunch with you."

What had gone through my mind in some flash instant was that several of us in Gordon's lab (Pat Jones, Bob Ivarie, Pat O'Farrell, Barry Polisky, who was in Brian McCarthy's lab) were close before Gordon's death, became even closer after Gordon's death, and wanted to continue to work together, if we possibly could. We realized no matter how fantastically good we thought we were, we knew the biochemistry department at San Francisco wasn't going to share that vision and wasn't going to offer us all jobs. One way for us to continue to have potluck meals together, play baseball on Saturdays, and go sailing was if we started an immunology farm on the northern Mendocino coast. We'd make antibody reagents and purify restriction enzymes. But, of course, we didn't have any money or business experience. Here's Ron Cape telling me he's the president of this company. I happened to meet with him a month later in Cambridge at a symposium on Science and Society: the Impact of Recombinant DNA.

I visited Cetus for the first time in late June. I gave a seminar on the work that Pat O'Farrell, Barry Polisky, and I had been doing at UCSF on expression of heterologous genes in *E. coli*. I was struck by two things: one, the total absence of any molecular biology equipment at Cetus, and two, the very sharp questions and interruptions I was getting from the nine or ten people attending the seminar. They were asking the questions and anticipating the data for the next slide before I showed the next slide. This had never happened to me at places where I'd given talks. There were generally few, if any, comments because it was crystal clear. In any event, that afternoon I was called and asked what it would take for me to come to Cetus and do genetic engineering. And I said it was impossible. He [Ron Cape] wanted to know why. I told him, "Well, first, I don't like the term 'genetic engineering.' It didn't take into account what's involved: one person doesn't do 'genetic engineering.' In addition, you don't have any of the space that's necessary, you don't have any of the equipment that's necessary, you don't have any of the facilities that are necessary. It's very expensive to set all this stuff up, and besides that's not what is important. What's important is who decides how things get done, who decides what gets done. People in industry don't have any understanding of what that takes. People who are making the decisions don't understand what's necessary, and it's just *no.*" And he said I was the most opinionated, biased person he'd ever met in his life, and it might be that way at other companies, but it's not that way at Cetus. He asked if I would put down on paper what I thought it would take to organize a recombinant molecular research lab. I had dinner with my friends that night, who encouraged me to put down what I thought was necessary. I said, "Well, that's ridiculous. They'll never accept it." They said, "That's right and your biases will be validated and you can say 'I gave it my best shot.'" Well, I didn't want to do it. So I made it as outrageous as I could think of making it.

I came back three weeks later from a meeting in Europe and a week after that Ron Cape called me and asked when could I start. I said, "Start *what?*" And he said, "Start the Recombinant Molecular Research Division at Cetus Corporation." I said, "Well, this is *not* what I was expecting." And he said, "Well, look, we considered it for the last month and the scientists believe it's a good idea, and Pete and I think it's a good idea, and Josh Lederberg and Don Glaser and Stan Cohen think it's a good idea, and that's exactly what we want to do and we want you to start it up." I said, "I don't know." I hadn't *thought* about it. He said, "Well, when you do think you'd be able to let us know?" and I said, "I have no *idea*. I don't know. And if that's not acceptable, I'm sorry." I just couldn't do it. "I don't know when I'll be able to let you know." And he said, "Well, I'll be calling you periodically." I had to think about this because it was serious. I talked with everyone whom I knew, who was doing what I thought I wanted to be doing when I grew up.

The people I respected said I should take the job at Cetus. In any event, the person I argued with most was Gordon Sato, because he had advised me to go to UC San Diego, he'd advised me to go to UC San Francisco rather than apply for jobs at Columbia or Texas or a postdoc with Paul Berg. He told me to do everything I ended up doing, and so I said, "Why are you telling me this?" And he said I was crazy. I said, "What I want to do after leaving Tomkins's lab is eventually to have a research group, have postdocs, and do the kinds of things you're doing. Why are you telling me that I don't want to do that?" He said that he spent more than 80 percent of his time trying to get money for graduate students, postdocs, technicians—and he had no time to interact with the postdocs and the graduate students. When he's on campus, it's faculty meetings. If I had any *reasonable* expectation that Cetus would be able to fulfill its commitment to me, don't hesitate for a moment. "Because," he said, "if you stayed at UCSF on your soft-money, nontenured research faculty position, you'll continue to work with people like Barry and Pat and other people, and you'll do nice things, but where will you be in five years?" And he said, "If Cetus is able to fulfill its commitment and you are able to attract the kind of people that you would like to hire and expand, there's no telling what the limits are." Since I had fallen in love with interactional science, largely through the five years in Gordon Tomkins's lab, I accepted the position at Cetus.

PR   David, if you were to change jobs, where would you go?

DHG   I don't know; I'm more thinking that it would be a biotech company than a university. I have great admiration and respect for academic science. The things that I was interested in doing, I would not be able to do as well in an academic setting as I would in a corporate position. *Provided,* what has always been most important, that there is a synergy of goals between what I, as a scientist, wanted to achieve and what the company wanted to achieve. And then, second, the nature of who your colleagues are. The two most important things are: Are you going to be able to do what's important for you to do? That requires a commonality of interests between your scientific goals and corporate goals. And second, who are you doing it with? Because 90 percent of what you do is collaborative.

It's very difficult, as an academic scientist, to do interactionist, collaborative science. The acculturation process is one that is keyed to individual, personal achievement. You first learn that as a graduate student. To get into the best lab as a first- or second-year graduate student, you'd better excel on the individual achievement scale to get through your orals and qualifying exam. In regard to your cohort class, you'd better excel in personal achievement as a graduate student. You have to prove that you can do independent experimental research. After that period of three and one-half to six years generally, one is applying to labs for postdoctoral study. Of course, if you're in biomedical or biochemical sciences, that is the most prestigious labs, i.e., usually the largest labs. You may be among the fortunate select few who have obtained a National Science Foundation or NIH grant or a Howard Hughes postdoctoral fellowship. You may go to the lab of Z, a famous Nobel Prize winner or about-to-be Nobel Prize winner, along with twenty other postdocs. Again, you have to— in the next two to two and one-half to three years maximum— produce high-quality publications in *Cell* or *Molecular and Cellular Biology* or *JBC* as independent publications. But you will, undoubtedly, being from a large group, be recruited along with several other postdocs who were coauthors with you on the most striking new transcription factor or growth receptor or some other hot field, interviewing for the same positions, competing with the same postdocs in the same labs. God forbid you were at UCSF in the mid-'70s and the postdoc in person X's lab competing on the same project with a postdoc in person Y's lab, on expression of insulin. Hardly an environment to foster interactive, collegial collaboration.

Through some miracle you may secure an assistant professorship at some prestigious place after this wonderful, illustrous career as a postdoc. Now you are competing, of course, for grants, and then renewals. Since all prestigious academic institutions have more assistant professor slots than associate professor slots, you are also competing with the excess assistant professors at your institution for a limited number of tenure positions. God forbid you should make the mistake as assistant professor of collaborating with either your former postdoc mentor—a very famous individual, of course, because you wouldn't have studied there had they not been—*or* graduate student advisor. All of your independent work done as an assistant professor will be ascribed to the brilliance of former major mentors, making it ever so much more difficult to get tenure. Thus one finds himself at age forty being promoted to associate professor with tenure and twenty years of experience of how *not* to collaborate. That one only gets ahead through individual personal achievement. But that's not the way I enjoy doing science.

Two

Cetus Corporation: A Credible Force

In March 1981 Cetus became a publicly traded company, a move that produced a major influx of desperately needed capital. In 1981–82 Cetus reorganized and redefined its objectives toward therapeutics, providing the commercial and scientific framework within which the company operated for almost a decade. Cetus was hardly alone in its enthusiasm for research on the interferons, interleukins, and other potential "magic bullets" for cancer or "keys" to the immune system. That enthusiasm was widely shared by many leading university-based scientists, industrial scientists, science journalists, various government agencies, and—last but far from least—venture capitalists and investors. In January 1983 a new president, Robert Fildes, was hired with a mandate to focus the company's numerous projects and make Cetus commercially successful.

State of the Whale (Cetus)

In 1980, in dire need of capital, Cetus management had turned to a consortium led by E. F. Hutton and Northern Trust Company of Chicago to privately place $50 million of stock. When that attempt failed, in March 1981 Cetus made a public offering of five million shares at twenty-three dollars a share. Although Genentech's highly publicized public offering was the first such offering for a biotechnology company, Cetus's was more spectacular yet, the largest single stock initial offering by a new corporation in U.S. history. The success of the offering was all the more remarkable given that Cetus openly claimed the company would not be profitable until 1985 at the earliest.[1] Money was abundant; biotech was hot.

The *Wall Street Journal* reported that Cetus's "preliminary prospectus, which features futuristic color photographs of scientists working with test tubes, disclosed that about one quarter of the company's shares will be in public hands after the offering is completed. Standard Oil Co. (Indiana) . . . will own 21.3% of the offering; Standard Oil Company of California . . . 17.3% and National Distillers & Chemical Corp. . . . 11%. The balance will be owned by Cetus employees and private investors."[2] The offering brochure listed risk factors, which included future operating losses. It also listed as risks Cetus's "reliance on unpatented technology and competition from genetic engineering and major pharmaceutical, energy, food and chemical companies. Cetus also said 'there can be no assurance' it will continue to attract top-flight scientists. . . . the company noted that its stock price may be highly volatile."[3] This note of caution was a formal disclaimer required by the Securities and Exchange Commission in a stock prospectus known to the trade as a "red herring." The investment community apparently assumed it wasn't true.

The successful stock offering set the stage for major expansion of both the staff and the physical plant. In the year ending July 1981 Cetus added 160 employees, bringing the total to 350 (of these fifty were Ph.D.s or M.D.s). As the first annual report phrased it: "Cetus management believes that the $107,216,700 net proceeds from our public offering in March 1981, plus the interest earned thereon, should be sufficient to support our programs until substantial earnings are realized from commercialization."[4] A state-of-the-art pilot fermentation plant for recombinant proteins was under construction in Emeryville (a small probusiness municipality located next to Berkeley), some old chemistry labs in the former Shell Oil building were being remodeled for expansion, and there was a dramatic increase in staff. Cetus's president, Peter Farley, was extraordinarily optimistic; he was a successful fund-raiser on a roll and felt that more money could easily be raised. During the second half of 1981 and throughout 1982 Cetus was negotiating with Japanese, American, and other corporations over investments for a variety of joint projects.

Cetus Corporation's first annual report, issued in July 1981, presented an upbeat picture of the company's commercial and scientific activities. The report told shareholders that the Cetus–Standard Oil collaboration on fructose production was on target with "integrated process finalization" expected in mid-1982. The yeast strain development and fermentation studies on ethanol, begun in 1979 as a joint venture with National Distillers, were going well. National had announced plans for construction of an ethanol plant having a capacity of fifty million gallons, to be completed in early 1983 at a cost of $100 million. The report lauded the progress of Cetus's well-funded and well-focused genetic engineering program for cellulose-degrading yeast. Work was proceeding on the new strains in an effort to demonstrate improved process economics, i.e., the ability to ferment cellulose. Negotiations were under way with a pulp and paper company; Cetus saw an opportunity to be at the head of the lignocellulose bioconversion field. Veterinary research using recombinant approaches to develop a vaccine for feline leukemia virus was proceeding satisfactorily. The contract with Roussel-Uclaf to improve vitamin B_{12} production had been successfully completed. Discussions were under way for similar work on vitamin C. Actually, these projects represented the company's bioindustrial past.

Cetus was also active in the more advanced recombinant technologies. In April 1981 a DNA-sequencing service lab was established to coordinate plasmid and phage preparation, oligonucleotide primer sequencing, and phenol distillation. The introduction of the sequencing lab was part of a reorganization and expansion of the program for bioactive peptides aimed at selecting and evaluating the most promising genes to clone as potential therapeutics. The first steps toward that goal included the isolation and characterization of RNA from cell lines and tissues expressing the activities of interest, and the construction of cDNA banks.[5] Research efforts were continuing on replacement peptides such as human growth hormones. Projects were under way in areas as diverse as monoclonal antibodies for possible use in transplantation operations and in the diagnosis of lower back pain. More important in terms of the company's future, "discussions with a producer of diagnostic systems are underway on the technical development and commercialization of a method for diagnosing disease at the nucleic acid level without the need for culturing often dangerous micro-organisms. . . . A broad strategy in the area of diagnostics is being developed in conjunction with a consultant."[6] Although this specific project did not turn out to be fruitful, diagnostics—and especially DNA probe technology—would unexpectedly play a large role in Cetus's future.

Becoming a Proven Threat: 1981–82

Despite the July report's optimistic portrait of the company's condition, the morale of the scientific staff was low, for interconnected reasons: there was a shared sense that there were too many projects under way, that there was chronic understaffing, and that the company lacked satisfactory mechanisms to choose between projects. During 1981 there was a growing sense, most keenly felt by those senior scientists running research and development, that more coordinated and directed, less ad hoc, procedures were required in order for Cetus to thrive. They felt that the company's top business management was insufficiently attentive to daily operations as well as delinquent in providing for focused research-and-development planning. There was no operative, centralized authority to arbitrate between the latest notions of the company's founders or Cetus scientists, and ideas coming from Cetus's subsidiaries and advisors. In the view of its senior scientists, Cetus was in desperate need of an overall manager for R&D.[7] The sense that a tighter rein was necessary was shared at many levels of the company. There was no consensus, however, about what form the company should take.

In April 1981, following the stock offering in March, Tom White was named head of Recombinant Molecular Research following Gelfand's recommendation to Cape and Farley. In an internal memo, White characterized the challenge facing Cetus in the following terms: "How do we improve quickly enough to become a credible force in biotechnology, and a proven threat to our competitors?"[8] How, in other words, was Cetus going to make the transition to becoming a top biotechnology company? White emphatically and repeatedly underscored the need for policy guidelines that would (1) produce a clear definition of who was running the company, (2) create a policy group with a clear mandate and real authority to provide the link between commercialization and research, and (3) introduce mechanisms for tracking policy issues and implementing a decision agenda. While these proposals sound like basic principles of business organization any M.B.A. would advocate, White's memos reflected and articulated the dismay and frustration felt by many of the scientists at Cetus. He hoped to improve the scientific climate while instituting procedures to evaluate individuals and laboratory groups closely with respect to creativity, productivity, and the ability to collaborate in research programs. The goal was a more efficient and stable environment by means of the formulation of a stable policy for initiating, planning, and carrying out new projects.[9] More routinized direction was felt to be the highest possible priority.

White introduced internal seminars that reinforced older disciplinary affiliations such as immunology, because he believed that interdisciplinarity and problem-orientation groupings had gone too far, and had deprived Cetus scientists of the more stable orientations that would keep them current on concepts and techniques in fast-changing fields. White took over the Shell/Cetus interferon project, which he felt had grown unwieldy with more than forty people directly involved. He instituted an in-house weekly newsletter, *Interferon Minutes,* in order to facilitate interlab communication. He hoped that this move would increase information flow and reduce the appearance of an ad hoc, arbitrary decision-making process. In a similar vein, White instituted a single search committee charged with all hiring at the scientist level, which centralized a formerly ad hoc process while giving research-and-development scientists a larger role in choosing other scientists.

White recorded in a memo his largely negative impressions of Cetus's annual scientific retreat in December 1981. In his view, the meeting's planners had been well intentioned but naive. The intent behind a series of problem-solving sessions may have been to obtain useful advice from the members of the scientific advisory board, but presentations had stressed technical difficulties and unsolved problems. Consequently, little time had been devoted to summarizing accomplishments or projects that were going well. In White's view, too much latitude had been given to the advisors and associates to promote their own projects instead of providing advice that would further the company's projects. Privately, the scientific advisors urged Cetus's top executives, Cape and Farley, to give more support to the immunotoxin and lymphokine projects. White and the other scientists were adamant: "No director can effectively manage the department or retain the respect of its scientists when he is ignorant of and uninvolved in decisions which affect its research. Unless this change is made by Cetus management, I will resign as director of Recombinant Molecular Research and return to my role as a research scientist or leave Cetus entirely."[10] White was clear under what conditions he could do his job professionally as he defined it. Finally, on 26 February 1982, a major reorganization was implemented: Jeff Price was appointed vice president and senior director of Research and Development; Tom White was appointed senior director of Research. They were given joint responsibility for all R&D in Cetus and its subsidiaries.

The second annual report echoes many of the points made in the first: during 1982 "we defined our concept of Cetus as a business. . . . We have chosen to place our primary emphasis on those applications of our science that we can realistically commercialize."[11] To accomplish this goal, a comprehensive business strategy would guide expansion of research facilities. Emphasis would be on "the development of products for which the proprietary technology developed by Cetus gives us a high probability of commercial success in selected markets."[12] These areas were diagnostics (chlamydia, maternal infections in early pregnancy), cancer therapeutics (interferons and monoclonal antibodies), and agriculture. The comparative weight of each area of research and development was still a matter of internal debate.

Diagnostics had a strong advocate in Kay Noel, who was appointed manager of Commercial Development, Health Care Products, in October 1981. Noel, who held a Ph.D. in biophysics from the University of Michigan, came to Cetus from Alpha Therapeutics Corporation (formerly Abbott Scientific Products Division) in Los Angeles, where she had risen through the ranks to become director of Marketing and New Products. Her responsibilities there had been primarily in the areas of business and marketing. Noel argued forcefully that diagnostics had distinct advantages over therapeutics: typically the product cycle was much shorter, regulatory hurdles were less daunting, and diagnostics, while often less glamorous, nonetheless had the potential to generate more revenue than most therapeutics. More specifically, she argued that Cetus had unique competitive technologies (DNA probes and monoclonal antibodies) that were good bets for commercial development if handled correctly.

In overall terms, White and Price agreed with these general objectives. There were major disagreements, however, as to how realistic Noel's assessment of specific projects was, given the time and resources allotted to them. They disagreed with Noel's position that the development of complex techniques such as DNA hybridization with nonradioactive probes and the identification of cancer-specific serum markers were mere "technical" tasks. Rather, they felt that a good deal more scientific work would be required before such products could be made commercial. In essence, the R&D scientists felt that Noel was selling unattainable goals to management. Her strategy, in part, consisted in presenting to management a picture of a sharp divide between what she characterized as Cetus's "academic researchers" and the more pragmatic and commercially minded people like herself who could solve problems. During the summer of 1982, Cetus management made the decision to devote more resources to diagnostic applications of biotechnology, incorporating important elements of Noel's approach into another fund-raising venture, the Cetus Limited Partnership Plan. Leaders of R&D felt that the targets proposed in the partnership were unrealistically optimistic; management assured them that the goals were flexible.

Over the course of the next two years, public conflict developed between Noel and senior scientists in R&D; good-faith consultation and trust broke down. Neither side had a monopoly on insight. As late as 1984, Price and White argued "that the DNA probe approach, while an important new technology, would never yield a format as simple and rapid as an immunoassay due to the intrinsic nature of nucleic acid hybridization.... the appropriate targets should be those not amenable to immunoassay based diagnostics."[13] The utterly decided tone of their memo underscores the degree of strain between the parties. Although their assessment reflected the then current wisdom, White and Price clearly misjudged the future scope of DNA-based diagnostics. Only a few years later, these same scientists would be centrally involved in developing PCR methods for infectious disease diagnostics.

Attempts were made by upper management to mediate the dispute, but as incident followed incident it became increasingly unlikely that the situation was going to improve. In this test of strength, Noel's days were numbered. Here, as in subsequent disputes, once the scientists of R&D came together on an issue and took a stand, it was difficult to ignore them because they were directing Cetus's research program, although in this instance, as in those that followed, considerable time and effort were often required to bring such disputes to a conclusion.

INTERVIEW: TOM WHITE

PAUL RABINOW   Let's start with the basic facts of your biography.

TOM WHITE   I was born in 1945. Both of my parents were scientists. My mother has a bachelor's degree in chemistry. My father was a doctor of science in bacteriology, and many of my earliest memories are of him bringing the experiments to the dinner table, where he apparently converted water into wine and back into water, and things of that nature. He would bring mice home from the laboratory, for my sisters and me to have as pets; and sometimes on the weekend he would take me to the laboratory.

We lived in Connecticut until I was about nine. My father worked for American Cyanamid. He was involved in the discovery of tetracyclines, and a drug for tuberculosis called ethambutol. Then, when I was nine, we moved to northern New Jersey, where he worked for Lederle Laboratories as director of Chemotherapy and Experimental Therapeutics.

From an early age I was interested in chemistry. I had a chemistry set. I made all the usual explosive things that children make, and followed that sort of inclination through high school, and went to Johns Hopkins University and got a bachelor's degree in synthetic organic chemistry and then came to Berkeley in 1967 and entered the biochemistry department. I left from the middle of '68 through the beginning of 1971 and was in the Peace Corps in West Africa.

PR   This is to get out of the draft?

T W   I was opposed to the war in Vietnam. I had looked into
leaving the country, and had gotten accepted at the University of
British Columbia, McGill, and Toronto, and then decided not to
take those options when I was accepted by the Peace Corps,
which was good for a year-by-year deferral, even though I was
in the 1-A category. So from roughly the age of twenty-two to
twenty-five, I was in the Peace Corps in West Africa. I trained in
Liberia and stayed on there for almost three years in a program
where I trained elementary school teachers how to teach mathematics as part of a team. Most of the teachers only had an elementary school education so we would teach them high school classes
at night. But there were certain fundamentally different ways of
looking at things that we learned while we were there. It was basically a very eye-opening experience for me because after some
amount of time I realized that the problem that the children had
with learning mathematics was that Western mathematics was
based on counting on base 10 and they counted on base 5. Therefore, they counted 1, 2, 3, 4, 5, and one 5 plus 1, one 5 plus 2, etc.,
and previous teachers had not realized how they counted in their
native language because these people hadn't learned how to
speak their native language, Loma. I think the other major effect
that my stay had on me was an appreciation for biology. We
lived in a small village in the rain forest, and the rain forest was
so spectacular with regard to insect diversity and biological diversity. One of the women in town was a biology teacher and I got
to know her and she was quite interesting. She was a real stimulus to me to get interested in the question of real biology—which
diminished my interest in chemistry.

   At the end of that I came back to graduate school. I was
twenty-five, and the government tried very hard to draft me. Actually, before passing my orals, I changed my thesis advisor to
Allan Wilson, an evolutionary biologist, because I was interested
at that stage in the evolutionary questions: where do we come
from? What are we doing here? and Where are we going?

   In order to be with a woman friend I went back and worked
as a technician at Johns Hopkins Medical School, which allowed
me, since I'd passed my qualifying exams, to take any course in
the medical school tuition-free. So I took primate anatomy and
pharmacology for a year. Then I returned to Wilson's lab, fin-

ished up in another year and a half or so, finishing my degree in 1975. Then I applied for a postdoctoral fellowship to study at the Department of Biochemistry at the University of Wisconsin with a molecular geneticist named Julian Davies.

P R   You were intending to be a professor at this point?

T W   Yes. Actually, I had mixed feelings about an academic career because during my graduate career I found the general arrogance of the professors in the biochemistry department to be practically intolerable, and the key feature of it that I found particularly irritating was that a scientist who had expertise in one narrow area of biochemistry felt by virtue of his position in the department that he was an expert on all aspects of life. In fact, many of them had led very restricted lives. I had been living for three years in the jungle and had traveled extensively in Africa and South America, and I had a different set of experiences and was not about to take instruction on how I should lead my life outside the laboratory from people who had no experience other than going from undergraduate to graduate school but nonetheless felt they were full authorities on all aspects of life. I had a general skepticism about both the medieval, serf-like quality of graduate-student life and actually also the treatment of anyone who had done anything other than go straight through your graduate career and then right on into the academic ladder.

P R   Was there a political dimension in this at all?

T W   Well, in 1967, in my first year in graduate school, in what seemed to be a relatively arbitrary way, the draft boards around the country allowed students to continue their graduate careers or they didn't. And then 50 percent of my class of eighteen students left in 1968. And only two of us returned. The other person returned the same year I did in 1971, and he dropped out before his orals, so it was quite a difficult adjustment process. For me it was not only hard to readjust to graduate school after missing it for three years, but it was hard to adjust to the U.S. because of having lived in a culture that emphasized personal values and the extended family and no value on material goods since everyone had the same amount of nothing.

And so I did receive an NIH grant for a three-year postdoctoral fellowship, but it was delayed because of the funding crunch. During that period, while I was waiting for the funds to become available at the University of Wisconsin, I painted houses in Berkeley and I made a hundred dollars a day, whereas I had made two hundred dollars a month throughout the previous five years. It was quite enjoyable. At the same time, I helped a professor at UCSF Medical Center apply for a grant to study the biochemical basis of breast cancer. I helped write sections of the grant and trained some people in the laboratory. It was very useful for me to learn the process of grant writing, although I didn't have that much confidence in the particular approach he was taking and tried to convince him to take other approaches.

This was the period of time during which I met my wife, a writer, Leslie Scalapino, and when the grant began at the University of Wisconsin, we moved to Wisconsin. I had a three-year grant but neither of us liked Wisconsin. Leslie left Wisconsin after about three weeks, which was a strong stimulus for me to find another position. So she would visit once a month or so, and I would come back and visit. I began to look for positions on the West Coast. On one of those visits I came to Cetus Corporation in Berkeley. I had known one of the founders, Don Glaser, from whom I had taken a graduate molecular biology course, and found him to be quite a stimulating teacher and interesting person. I was also impressed by the scientists that I met at Cetus. As I said, it was a time of few openings for scientists, and after having gone to Berkeley, and Wisconsin, and Johns Hopkins, I felt I deserved a better position than Northwestern or the University of Illinois for a job. Cetus offered me a position and I took it in 1978.

PR   That must have been a very bold move?

TW   Right. I think it was at that time, and I was criticized by some of my colleagues for taking that step. In fact, one of the scientists at Wisconsin, a famous evolutionary biologist named Walter Fitch, said: "Well, you are obviously not going to do any good science from now on"—that sort of thing, and tried to talk me out of it. In contrast, Julian Davies, who was the head of the lab, had industrial scientists come for six months sometimes and during the summer. I was impressed by the industrial scientists who had come to Davies's lab. So it made me think that it was really more an academic preconceived notion that the work in a company was second-rate. At that time, there wasn't still really that much publication from companies, so it was difficult to know and, therefore, easier to maintain your prejudice. In any case, after having given a number of seminars at scientific conferences and at universities as part of the job-hunting procedure, the most astute questions that were asked me by any of the people at any of the seminars I had given were, in fact, from the people at Cetus, who were an unconventional sort of rambunctious group, and really seemed to be in a position to have the resources available to make the most use of their talents, unlike the struggle in the academy.

P R   So, entering Cetus in the late seventies was not really the same thing as going to work for Du Pont.

T W   Not at all. In fact, Cetus was the *only* biotechnology company. I think Genentech either hadn't been founded or it was about to be founded—and Biogen had not been founded. So it was considered very unusual to do this, and Cetus at that time had just begun to form a recombinant group which was only one year old because the techniques were only two years old.

P R   Describe the atmosphere.

T W   Well, at the time I joined I still wasn't sure exactly what I would work on. They were under contract with a large French pharmaceutical company, Roussel-Uclaf, to do the vitamin B_{12} work. That particular project was confidential because there were only two or three producers of vitamin B_{12} in the world. The competition over the production of this vitamin as a commodity was very intense and so the project was sort of secret.

P R   One of the stigmas of industry work is that it is secret. In this case, then, it is true.

T W   Right. At the same time I began to work on a veterinary vaccine—that work wasn't secret. I proposed some new ways of approaching the problems and was basically told, "Well, if you think you can do it then tell us what you need." So, though I had been hired for one thing, I proposed a totally different approach to something else.

PR   Who said, "You can do it"?

TW   My colleagues and also one of the more senior scientists who served as the head of the B$_{12}$ project, Jeff Price, who ultimately became the head of Research and Development at Cetus. I think one of his strengths was that he appreciated innovation as well as an oddball approach to a thing; that is, something that was really outside of a person's formal expertise which prevented you from becoming categorized as being a biochemist or whatever. So, in fact, my training, since it was rather broad, unconventional really, in a strictly academic sense, was appreciated there because when they needed something from immunology I often knew how to do that and no one else did. They wanted something synthesied, I could do that, so actually that began to make me feel like, well, this place really appreciates someone who can make use of resources that are beyond what they hired you to do. That approach allowed me to have a broad effect on a few projects, and ultimately in 1981 Cape and Farley asked me to become the head of the molecular biology department partly because in meetings when project technical difficulties were being discussed, I sometimes would have the solution in the context of our effort. So they began to think that I knew how to do the right experiment and also could propose that experiment in a way that was not inducing conflict. In some projects, there were frequent conflicts, and people would stop talking to each other and go off and not work as a team. So at some point they asked me to head the department and change things around, which I did. I didn't want to be the head of the department and I argued forcefully to not take on that role. The founders of the company insisted that I take it on or propose some alternative, and I proposed several alternatives and they didn't accept them, so I agreed to take it on with the caveat that if I didn't do any good, that I could go back to what I knew I was good at, which was laboratory science. Secondly, I insisted that I would be able to retain a technician to follow my own academic interests, which were in the field of molecular evolution. They agreed.

P R   Even though it had no commercial payoff?

T W   No relationship to the commercial interests of the company whatsoever, which was another thing I sort of appreciated and felt was a fair trade. I had an area that was completely independent of any of the company's projects, though the company's projects interested me because they were cloning genes that had never been cloned before. And it was the very unique characteristic of getting scientists from very different disciplines like biochemistry, molecular biology, and cell biology working together in an industrial setting that was quite different from the academic model, where people have their own labs and they set their own direction. The chairman of the [university] department does not direct the individual investigators nor are they compelled in any sense to work together as a group. If they choose to collaborate they may, but rarely. Especially in those days, it wasn't as frequent to have the larger groups of collaborators that you find today. The younger scientists had a strong say in policy. The situation seemed perfectly normal to us. Why shouldn't it be like this? Yet we had the sense that it was special too.

SCIENTIFIC FOCUS: CANCER THERAPEUTICS

At the end of the 1970s, researchers in molecular biology, immunology, and related fields began to develop additional strategies, moving away from known molecules, such as insulin or growth hormone, to focus on unexplored domains, the biological molecules that seemed to have important regulatory roles but whose precise function, or range of functions, remained unknown. This new focus was above all on the immune system, particularly on the interleukins—proteins secreted by one component of the immune system that stimulate other components of the immune system into activity.[14] If this activation of the components could be artificially induced, then it might be possible to regulate the body's immune response. If the immune response could be crafted to respond to specific antigens, stimulated or moderated upon command, then a truly revolutionary therapeutic tool would be at hand. If the immune system could be approached as an autonomous system and then regulated, a radically new era of medicine, science, and commerce would be possible.

Cetus and other biotechnology and pharmaceutical companies wagered that the emerging technical mastery of genetic matter *in vitro* would be rapidly translatable into therapeutic products effective *in vivo*. This self-reinforcing trend led business management, scientists, and venture capitalists to center their scientific program around potential therapeutic products and to aim their corporate strategies at winning the race to discover and market the promised potent molecules. In 1979 Cetus had begun research on one promising immune regulator, the interferons, that led to a joint agreement with Shell Oil Company on 1 July 1980. Cetus took responsibility for the scientific work and Shell for product development (regulatory affairs, clinical programs, and market research). Some internal debate was carried on at Cetus about the wisdom of infusing so many resources into the interferons when other companies already had a significant head start. Some Cetus scientists thought it was worth casting their net more widely, to explore other molecules from the immune system. The enthusiastic consensus at Cetus, however, as in biomedical research in general, was that the interferons had potential clinical value as antiviral drugs. Why were interferons considered such an exciting class of substances?

In the mid-1950s a small group of researchers was trying to figure out why animals rarely seemed to suffer from more than one viral disease at a time. They discovered that when a cell is invaded by a virus, it produces a protein that provides a short-lived immunity to other viruses. In 1957, researchers identified the protein that seemed to "interfere" with viral infections; it was given the name interferon. Although research on the protein's properties continued for the next decade and a half, progress was constrained because the protein seemed to occur naturally in such small amounts that it was exceedingly difficult to isolate or produce. Because of the laborious process of isolating it, a gram of interferon would have sold for $50 million, making it very expensive to analyze properly and prohibitive to use for any therapeutic purposes. The science of interferons languished.[15]

By the mid-1970s, however, inflated claims for the family of interferons were beginning to gain broader currency. During the heady days of Richard Nixon's "war on cancer," when hopes were high and money flowed, interferon was touted as the most promising miracle drug, the potential cancer cure. The idea that interferon might be useful in treating cancer can be traced to the belief, which held increasing appeal at the time, that certain viruses were one of the probable causes of cancer. Although at the time there was no proof, experimental or otherwise, that viruses caused cancer in human beings, there was clear experimental evidence of cancer viruses in animals. A half century earlier, the Rous sarcoma virus had been shown to cause cancer in chickens. A plausible logic was constructed: since viruses caused cancer in some animals, and interferon interfered with viruses, interferon could well be a miracle drug to prevent or cure cancer.[16]

In 1979 the first human interferon gene was cloned[17] and during the next eighteen months more successes followed. Giant multinational pharmaceutical companies joined with start-up biotechnology companies that were staffed with the proverbial hardworking, talented, ambitious Ph.D.s and backed by the symbolic capital of a new breed of bioscience stars. The Swiss company, Hoffmann-La Roche, established a collaboration with Genentech to work on alpha interferon. Schering-Plough teamed up with Biogen; on 16 January 1980 Biogen held a press conference in Boston, hosted by Harvard's newly Nobeled Walter Gilbert, to announce the production of interferon in bacteria. By 31 March 1980 interferon was on the cover of *Time* magazine. Much of the research on interferon structure-activity relationships and on defining its range of biological activity shifted to the private sector. As science journalist Robert Teitelman aptly observed:

> Interferon was a dress rehearsal for bio-technology. . . . It created a space for entrepreneurs in academia and the bureaucracy, and it left a sense that a revolution was . . . not only necessary but possible. It also left in its wake the lesson that image meant more than substance, that means could be compromised to achieve desirable ends, and that

a certain scrupulousness could be abandoned for expedience. All these were notions that dovetailed nicely with certain aspects of Wall Street.[18]

Teitelman's use of a dress rehearsal as a metaphor for the interferon phenomenon, however, is misleading because it implies that the script and the actors already existed. It also suggests, once again, that the previous theater had been a pastoral—one in which image, expediency, and compromise had had no role. Be that as it may, Cetus certainly was an improvisational space.

Collaboration

Cetus, in partnership with Shell, was continuing its interferon projects. However, by the end of the 1970s most of the interferon genes had been cloned. White and other Cetus scientists felt it was important to diversify the research agenda if Cetus was to remain competitive and proposed expanding their collaboration to the "third" category of molecules, the so-called bioactive peptides or immune modifiers. After a long hesitation, Shell declined to participate. Energetically pushed by scientists at its Palo Alto–based subsidiary, Cetus Immune, Cetus decided to go ahead on its own and placed interleukin 2 (IL-2) at the top of its bioactive peptide list. At about the same time (fall 1981), Steven Rosenberg and his colleagues at the National Cancer Institute were reporting promising results using crude interleukin 2 prepared from human cell lines. These results indicated that IL-2 might well play a role in strengthening the immune system and might be a potent anticancer agent. IL-2 was becoming the "molecule of choice."

The first obstacle facing researchers was securing quantities of interleukin sufficient for carrying out experiments. The standard method, extracting IL-2 from massive numbers of spleens taken from mice, was expensive, inefficient, and yielded insufficient quantities. Although attempts to industrialize the procedure were under way and some increases in yield were achieved, the lack of a ready supply of IL-2 remained a source of constant frustration. The shortage was felt even more acutely when studies of laboratory mice seemed to confirm the validity of an experimental approach developed by Rosenberg at the National Cancer Institute. He writes:

In an experiment in late 1981, we achieved our first cure. Every untreated animal died within twenty-four hours after injection of tumor cells. But with 50 million cultured and sensitized T cells, 80 percent of the treated animals—animals with large, established tumors as well as disseminated metastases—survived indefinitely. T cells had found and killed the [cancerous] cells everywhere in the animal.[19]

Rosenberg's accomplishment rapidly spurred efforts to modify and refine delivery systems and treatment protocols. It enticed more researchers and more institutions to this area of research. It gave a sense of impending triumph over cancer that authorized a "rigor" in therapeutic regimes that tested the limits of ethical practice.

Rosenberg's quest for an ample supply of IL-2 brought him to Cetus. He accepted an invitation to give a talk on IL-2 at the Cetus subsidiary, Cetus Immune Corporation (CIC), in Palo Alto, on 30 July 1982. His talk was well received and he returned for a follow-up talk three months later. While Cetus scientists intended to begin clinical trials for beta interferon as soon as possible, they were encountering delays; nonetheless, moving beta interferon into clinical trials remained at the top of the 1983 Cetus project list. Any new treatment must go through a series of processes designed to demonstrate that it is effective and to define what harmful side effects it may produce. The clinical trials of a drug or treatment in humans must be approved; to seek approval researchers must prepare an Investigational New Drug Application (INDA) to begin Phase I clinical trials. Such proposals require extensive paperwork, a standardized manufacturing process, and coordination of data from earlier studies, as well as frequent interaction with government officials. Concurrently, cloning work on IL-2 was progressing at Cetus and elsewhere. Rosenberg was moving ahead with his own proposal to initiate clinical trials at the NIH with IL-2 supplied by Du Pont. While Cetus was pleased that progress was being made, the relationship between Rosenberg's trials and its own remained unclarified. Cetus wanted to obtain proprietary protection for therapeutic uses of IL-2 discovered during outside collaborations.

Cetus and Rosenberg had major strategic and ethical reservations about each other. Rosenberg was concerned by the possibility
of encountering obstacles due to patent or confidentiality arrangements. Some Cetus scientists worried about the effect on Cetus's
reputation, and hence on its commercial future, of too close an
association with perceived scientific and ethical recklessness. Rosenberg was well known for pushing clinical trials to the limit.
However, both parties felt that they needed the other. In 1982 the
Cetus subsidiary CIC approached Rosenberg about joining the
company. He refused:

> The emergence of the biotechnology industry was con-
> stantly creating conflicts . . . between the traditional free
> exchange of scientific information and the commercial ex-
> ploitation of that information. At that time, people in
> other labs began refusing to send me reagents unless I
> signed a confidentiality statement, which I have never
> agreed to do. . . . Science has always been part of the world
> and never wholly innocent. But it had been a special part
> of the world. It was losing some of its specialness. . . . it
> taught me never to be naive in my dealings with a bio-
> tech company.[20]

Rosenberg neglects to mention that the confidentiality agreements
included those from university and government labs. Rosenberg
was collaborating with Robert Gallo at the National Cancer Institute. Among the other non-"traditional" actions attributed to
Gallo is his accelerating a trend toward *not* sharing materials and
information with his colleagues.[21] Although Gallo's behavior
might be linked to the fact that he was among the first to benefit
financially from the new regulations permitting government employees to earn patent royalties, his motivation may have had more
to do with the race for recognition as the discoverer of the cause
of AIDS. The competitive environment was inhabited by scientists
employed by diverse institutions.

At Cetus both Rosenberg and IL-2 were viewed with a certain
caution, the result of scientific, commercial, and ethical doubts.
There was a clear sense that although there were good scientific
reasons to bet on the importance of IL-2, it was far from certain

that IL-2 would be the type of miracle cure Rosenberg believed it would be. Finding a cure for a specific set of well-understood diseases seemed likely to be a long process. Regardless of scientific potential, the federal regulatory environment made drug development a gamble and a long and costly process. Further, and somewhat paradoxically, the effort to produce a natural molecule rather than develop an artificial one could well add an extra degree of complexity and difficulty, because natural substances were likely to have evolved to play complex roles in disease prevention. Exploring these roles might well prove to be a protracted undertaking.

For these and related reasons, Cetus wanted to make sure it provided a pure product, one that would express only the desired qualities. Achieving that purity took time. Cetus's team was working seventy-hour weeks; the company had a lot to lose by precipitously releasing its product. Rosenberg was in a hurry—in the view of some at Cetus, dangerously so. There were scientific and commercial reasons for securing more information from as many collaborators as possible. The widest possible collaboration could be an industrial strategy as well as an academic one. Toward that end, Cetus was beginning to send IL-2 to what would eventually total more than one thousand collaborators, the majority of whom were performing more traditional cell-culture and animal studies. Some at Cetus felt that they should wait to see what these studies showed before entering into high-visibility and high-risk clinical trials. Others considered the substance so promising that it should be given the highest priority and pushed vigorously. Rosenberg's drive and high visibility were double-edged swords. On the one hand, he was well placed enough within the government bureaucracy to get his studies approved rapidly. On the other hand, he had his own agenda and would be unlikely to have any loyalty to Cetus. Further, "he had a reputation as a cowboy and might do something dangerous with the material."[22] Results from both Rosenberg's own work and the work of others were raising troubling questions about IL-2's toxicity. Everyone agreed that risk was an inherent part of experimentation, but sobering questions were raised about how much preliminary testing was necessary. It was clear to Cetus that Rosenberg would be very aggressive with IL-2.

TRANSITION

By mid-1982 Peter Farley's 1978 claim—"We're building a new IBM here. . . . we see ourselves as far and away the number one company . . . applying modern biology to industry"—was sounding hollow.[23] While the National Distillers project to produce ethanol was technically successful, demand for gasohol was so low that long-range scale-up plans were scrapped. In May 1982 Standard Oil of California pulled out of the fructose venture, an event that focused Cetus's attention on the drain that overhiring was having on the funds gained from the stock offering. It thus provided a convenient excuse for the layoff of eighty-nine employees within six months. The venture capitalists and other outside interests now financing Cetus pressured management to change. As the *Wall Street Journal* commented: "The withdrawal of Socal, which owns 17% of Cetus, appeared to be a major blow. . . . In the first nine months of Cetus' current fiscal year, fructose related revenue of $1.6 million accounted for about 8% of Cetus' total revenue. Perhaps more significantly, the contract with the major oil company contributed about 18% of Cetus' research and development revenue in the first nine months."[24] By the summer of 1982, it was clear to all concerned that Farley and Cape needed help to manage Cetus, and the search for a new president began. The *Wall Street Journal* reported that "Cetus Corp. said it is narrowing its 'smorgasbord' of scientific projects to concentrate on commercially feasible products. . . . Analysts, while generally applauding Cetus's strategy of tightening up, are waiting to see whether the company can bring it off."[25]

During the fall of 1982, Cetus interviewed a substantial number of candidates for the position of president, all proposed by a New York "head-hunting" firm. As it happened, Robert Fildes had been interviewed and rejected for the same position at Cetus a year earlier. Insiders remember Cape, a cultivated and soft-spoken man, reacting negatively to Fildes's aggressive demeanor and frequent use of the vernacular, especially expletives. Some observers feel that as Fildes's position at Biogen worsened, he may have moderated his style in encounters with Cape and won him over. Of course, most crucially, Cetus's position was declining and its need for a "strong" leader became acute. Fildes's mandate was to take charge of the daily operations and direction of the company, to give Cetus a unified mission, to streamline it, and to make it profitable. In order to achieve these goals, he developed a strategy to make Cetus a full-fledged biotechnological therapeutics company.

INTERVIEW: ROBERT FILDES

PAUL RABINOW   Can you tell me about your background?

ROBERT FILDES   I was born in England in 1938, where I grew up. I went to college there. I spent all my time, actually, at the University of London, at one or two different colleges. I finished up getting a Ph.D. in biochemical genetics. I spent a couple of years doing postgraduate work and then decided that I would go into industrial research.

PR   As a youth, were you interested in science?

RF   Yes, I think I found doing science easy compared to doing some other things. And I think when you find something that interests you and that you can do reasonably well, then you tend to orientate towards that. The decision about what I wanted to do with my life was very much my own, because from the age of about sixteen—my father died when I was quite young, and I left home when I was about seventeen. I went through a sort of somewhat different process to a typical situation. The typical situation in England is, at sixteen you do the certificates, and you do three [certificates] to the advanced level, then you go off to college and move on to a Ph.D. In my case, I didn't have the opportunity to go on and do the final two years of advanced schooling. Between the ages of sixteen and twenty, I did the advanced requirements to get in a university on a part-time basis, in the evenings, and worked during the day. Eventually, when I got the requirements to go back to the university, I went back.

In that period away, I worked for a pharmaceutical company. I was a young technician, you know. I used to clean out animal cages, those "low totem pole" jobs. There were several people who I worked for, scientific staff at the company, who gave me encouragement and advice in regards to what to be studying in

the evenings. So I had a lot of marvelous support from those
folks. And I'm sure that that helped tremendously, because you
need encouragement.

PR   So it was a supportive environment?

RF   Yes, and these were successful scientific people. Most had
gone through the university system and got Ph.D.s and were
now doing research in industry, and I worked for them and they
gave me that sort of encouragement. Then, at twenty, I went
back to the university, got a scholarship from the government.

My advisor for the Ph.D. was a guy called Professor Harry
Harris, and he was the head of the Medical Research Council.
It's a bit like the NIH here, I guess. It sets up special groups to
carry out research. The groups don't have teaching responsibilities. Harry was, at that time, considered to be one of the leading
geneticists, and he had a big influence on what I did with my
postdoctoral.

PR   At this point were you thinking you were going back to industry?

RF   No, in my mid-twenties I had concluded that I liked doing
basic research and I liked the academic life. It seemed a very com-
fortable existence. I got drawn back into the industry quite by
chance. My wife was working for Glaxo Laboratories, a big pharmaceutical company in Britain. She was the secretary to the head
of Human Resources of that company, who had just joined the
company and had relocated from the north of England to the
south of England. My wife said to me one day, "You know, the
man's down here alone. His family's up in the north. Why don't
we have him over for dinner?" He came over, and after dinner,
you know, he got talking about careers, and he said, "Have you
ever thought about going back into industry?" And I said, "No,
not really." He said, "There's always lots of good opportunities
for people in this industry." And that sort of began a process that
took place over two or three months where he actually, at one
stage, said, "Why don't you come and meet the head of Research
at Glaxo?" So unlike, I guess, the typical interview process, I actually was invited to lunch at the company with the head of Research. To my surprise, a week later I got an offer from them to

join them as a senior scientist. He painted a picture which was very attractive to me.

P R   Was there a really sharp line in England between the academic and industrial worlds?

R F   Yes. I always remember when I went in and told Harris that I was leaving, giving up my postdoctoral position with the MRC. He looked at me, and he accused me of prostituting. We're back in 1966, something like that. Yeah, generally speaking, I think folks who did academic research were different and maybe superior to folks who went into industrial research, in their own eyes. Of course, that has melted away over the years.

P R   What was the attraction of Glaxo?

R F   I think one of the attractions of going back into industrial research in the pharmaceutical industry was to be able to work on science that could have an impact on how you treat a disease. When I think about it, I decided that pursuing science for knowledge alone was not going to keep me happy as an individual. It had to be something more.

P R   What were your initial duties, and where were you working?

R F   Well, initially, as I say, I went in by the back door to some degree. I went in as what was called a senior scientist in a big company. The lowest of the senior scientist grades. In that position I had a lab, and I had two assistants, and I was part of a section that was called Industrial Biochemistry, and I think in the whole section there were probably seven or eight people like me with similar group sizes. One task was to develop alternative ways for making some antibiotics that were being used to fight infectious disease, and the other one was to seek new product opportunities, in particular, in the field of industrial enzymes.

P R   So the enzyme research was really exploratory?

R F   You see, you have to understand Glaxo. It had two types of production capability. One was chemical synthesis. The other was fermentation. They had enormous fermentation capacity. You're talking about tens of thousands of liters. They were using those to produce a variety of natural antibiotics, but they were also exploring what other types of products of industrial importance could come out of fermentation. It was just at the beginning of the idea of using enzymes, for example, in washing powders. So, for example, people talk about, "Well, maybe they'll come out with an enzymatic treatment for shaving." Oh, we found lots of enzymes that would take hair off our faces. The trouble was it would take the skin off.

P R   Small problem.

R F   Small problem. Basically we were looking for new products that could be made by fermentation that could ultimately take up some of these enormous production capabilities that we had. You couldn't assume that you were going to be making penicillin forever, or you were going to be making vitamin B_{12} forever. So part of the task, my task, was to try and identify some new product opportunities that would fit into our existing production capabilities.

P R   For how long did you do that?

R F   I was with Glaxo for nine years. I was in that position for about two years. Then they promoted me to take over the whole section, which was six or seven other groups like my own.

P R   So you were moving up quickly? Does this take you out of the lab?

R F   It did, but at that stage now, what you find is that you've got like six or seven Ph.D.s reporting to you, and each has got its team. Instead of directing individual technicians and assistants to do experiments for you, you're planning, now you're really working with scientists who are actually doing that, and all you're trying to do is to coordinate those efforts appropriately and to manage those efforts to make certain that people are focusing on the things that are appropriate for whatever the company's interests are and to provide them with whatever support they need. So it becomes much more of a management role, although, obviously, you're still close enough that you're having little discussions about the science with each group, because what it does is it exposes you to a whole spectrum of different scientific projects, which

I found fascinating. Instead of working on two different things, now I'm involved in talking science about ten different things.

After about a year, they promoted me again. And then they put me in charge of development. The first two years, the first two positions were really basic research, looking for new products. Now, I had also this development responsibility. I also got exposed then to nonbiological research, because a lot of the process side is chemistry. What we were trying to do was to find what kind of processes to make products, the kind we already had, but to make them more efficiently. So I did that for a couple of years, and then they promoted me again. At that point, I moved up to their biggest production site in the north of England, certainly one of the largest fermentation sites in the world. They made a whole range of fermentation products and chemical products for drugs and treatment of infectious disease, vitamins.

I was there for five years. Obviously they felt my handling of people, and the groups, and so on, was something that they felt was a positive thing because they kept pushing me with more and more challenges. It's great fun working with people. So it has always been interesting to me to put together groups of people, especially clever people, and see what we could accomplish as a group.

PR Quite early in your professional life you've been in leadership positions. What happens next?

RF By 1974 I was head of fermentation R&D, which was a fairly substantial group in the parent company. I also began to represent the company in terms of going abroad, handling meetings, and looking for technology opportunities. And I came to America quite a bit during the '70s, and I got to know it. About that time, I started to get approached by several American companies, I guess based upon my meetings and discussions, and my presence at meetings. I always found America very fascinating in those days, and clearly there was a lot less structure to things over here in terms of one's ability to move up, although, curiously enough, I hadn't, up until then, encountered any obstacles at Glaxo. They had been very good to me. But maybe I've always been one who was sort of curious about things on a broader base, and I certainly was interested in international things. And so when American companies started approaching, I began to think about it.

PR   Were there any class or political considerations that came into play in terms of your position in society, or ceilings to what you could see yourself doing, in England?

RF   Yeah, I think there was. I mean, I guess there's always that concern. You know, I came from an ordinary working-class family. As I told you, I pretty much had to work my way out on my own merits. Fortunately, it was a time in England when it was possible for people to do what I did. There was a time in which the Labour government had set up a grant scholarship system, which was open to everybody. Because of that environment, I was able to progress, even though it was somewhat unorthodox. I guess there was always in the back of my mind—well, even today in England there is still a class structure. I don't care what people say. I guess there was always in the back of my mind whether or not I would ever be able to move to the very highest levels of management in Britain. Obviously, I was ambitious. I was beginning to sense that maybe there was going to be a cap. The Americans were making me all sorts of incredible offers of opportunities that I might never see in Britain

I took the leap. Bristol-Myers had decided to reorganize a major part of their activities, and to do this they set up a new division called the Industrial Division. The purpose of the Industrial Division was to pull together all of the production activities in the major places in the world and coordinate all their production capability and service the various market divisions. And they also built a new management team. I was vice president of Development, and my job was to coordinate the technical activities within this division. So it was an opportunity to come in on the ground floor and help create a new organization.

PR   So once again, you're in a good position, you're appreciated, you're enjoying the work, and so what happens next?

RF   I was with Bristol-Myers from '75 to 1980. Then in about 1980, I get a call from a venture capitalist type guy, Moshe Alafi.

He says, "I'm in Kennedy Airport. I'm just about to fly to Geneva." He says, "There's a great opportunity I'd like you to think about. Have you ever heard of a company called Biogen? I'm on the board of that company. We're going to set up a U.S. division. Would you consider coming on board and doing it for us? Everything's in Geneva, and we want to have a U.S. equivalent, and we need somebody with your sort of experience to literally build the company from scratch for us." And I said, "You're joking." He said, "No, no, no. It's a great opportunity."

To make a long story short, we talked about it, and I met some of the people involved in it, and I was impressed with the scientific credentials of those people and what they were trying to do. And again, it wasn't that I was unhappy at Bristol. In fact, I had been promoted. I was on all their special programs and clearly well thought of. I got the bug. I mean, here's a chance to build something from scratch. Here's a chance to be on the cutting edge of science in terms of maybe to be involved in some pretty exciting breakthroughs and a chance to make a lot of money. I realized, go for it.

PR   Was this a high-risk change of position?

RF   You know, it's overplayed, this risk. If you're good at what you do, you shouldn't worry about—if it succeeds, great. If it doesn't, go off and do something else. But if you've always got the confidence, if you've always got the knowledge, if you've always got the ability, you'll find another job. I never even thought of the risk thing.

PR   Did you talk to Wally Gilbert before joining? The story is that Gilbert proved to be pretty difficult to get along with.

RF   Well, Wally's a fascinating guy. With the sort of career I've had and the organizations I worked in, big organizations, you see a lot of manipulations, and you see a lot of the power plays. You know, people trying to get ahead by knifing their colleagues. Maliciously carrying out things in order to benefit themselves. That wasn't Wally at all. Wally was difficult because Wally was naive about people, about anything outside of science. I mean, he was like taking to a very bright child and putting him in a cookie shop. His reaction was he wanted to try all the cookies. He had no professional skills outside science, so he knew nothing about product development, nothing about manufacturing, but he thought because of his intellect, "Well, that's a snap." You know, he could do that in a day. I used to joke, "Put this one in a padded cell. Don't lose him." I guess the one thing I learned from working at Biogen, and working with so many brilliant scientific people, some of them completely uncontrollable, is that you have to find some way to control their influence outside of the science, because if you let them run wild in every direction, then you have an unmanageable situation. At the same time, let's have an organization which encourages the scientist, the production guys, the regulatory guys to have their say about it, but ultimately it's got to be the marketing guy who makes this decision if it's a marketing matter. Encourage people to have input into what you're doing, but keep the final decision in the hands of the appropriate experts. If it's finance, finance. If it's marketing, marketing. If it's research, research.

I spent the two years at Biogen, and then, because I got to the point where I was totally frustrated with this totally disorganized approach. So I moved on to Cetus.

Ron Cape and I were founding members of the Industrial Biotechnology Association, formed some time in the early '80s. I was in a meeting with Ron and I said, "I just read somewhere that you're going to set up a subsidiary in Britain. You know, if you're looking for somebody to do it, I might be interested." He said, "Well, wait a minute." So later on at the end of the day, he said, "You know, if you're really serious about going, we're in the process of trying to bring somebody in on Cetus, and we should talk about that." Eventually we got to the point where I wanted to join.

P R   How do you see Cetus at this point?

R F   Ron bought into the idea that Cetus needed now to be in focus, needed to build the capability to go beyond research, and that required a different management. When I agreed to join them, one of the things that, obviously, I wanted to know from them as clearly as—you know, there's always this question about being founders—we talked a lot about what was going to be my role and what was going to be theirs. I was going to have the ability to build a company, and they would not interfere. And it was agreed that we wanted the company to be a successful pharmaceutical company.

Cetus was an incredible science machine. Lots of exciting things were going on in terms of creating leads. So we had to decide where we were going to take this thing. The question was what to make. You've got to remember that back in those days Cetus was being described as this sort of university, meaning here is a group of people who are doing lots of interesting research, but there's no commercial direction there. They had just gone public and raised more money than anybody else had ever done before. And they had been through a very difficult time, too, because once they had gotten this money, they then went on a totally crazy recruiting binge where they recruited lots of people without even knowing what they were going to do with them. And then they had a layoff. In fact, they had a layoff about three months before I actually joined.

PR So there was a real basic harmony of views?

RF Well, there was a harmony of views between myself, and Ron, and the board of directors. I'm not sure that there was a harmony of views amongst individual scientists in the company. Obviously, there was a lot of internal competition for attention and support. The emphasis was clearly going to go on what we'd do in the therapeutic area. We had IL-2, we had interferon, we were working on several other things, so we set about sorting out the priority amongst those products, opportunities, and deciding what we had to do to take them forward and find whether they worked or not. We began to focus the thinking, the planning, the execution, and the support of that therapeutic job.

PR What was your horizon for the therapeutics, in terms of becoming profitable?

RF We were thinking that we could do these things in about five years, which turned out to be too ambitious.

FOCUS: MUTEIN TRIALS

Once at Cetus, Fildes expanded the management staff, hiring senior executives from Bristol-Myers in operations, production, and marketing, and a patent-law expert from Exxon. Fildes immediately provided what he had been hired to provide, strong direction to the company. As the 1983 annual report told stockholders, "Cetus intends to be a totally integrated human health care company in North America."[26] Clearly, the plan for such an operation would have to be carefully but boldly charted. Fildes chose to focus on cancer: "we are aggressively concentrating our development efforts on cancer diagnostics and therapeutics and infectious disease diagnostics."[27] This was a high-risk, high-benefit strategy. Cetus moved rapidly to implement Fildes's plan; within the first fiscal year after Fildes's appointment, more than 70 percent of Cetus's efforts and resources were devoted to this aim.[28] On 10 May 1983 a *Wall Street Journal* headline reported: "Cetus Vice-Chairman Farley Quits to Develop New Firm." "'It's all quite amicable,' he said of his departure, explaining that Cetus had completed its transition from an entrepreneurial company with the appointment of a seasoned manager to handle its day-to-day operations."[29]

Although subsequently there would be the creation and dissemination of a myth that the "beached whale" had been saved by the firm hand and keen business sense of its new, aggressive, businesswise CEO, Robert A. Fildes, the process was, in fact, well under way when Fildes came on board. The myth was convenient for the repair of Cetus's image as well as for Fildes's reputation. In sum, a new structure was emerging during this period, one that remained largely intact in subsequent years. In the latter part of 1982, the final lap of the race to clone IL-2 had clearly begun. Cetus, Immunex, and Genentech were about to lose that race to the Japanese researcher Tadatsugu Taniguchi, who isolated the gene for IL-2 in late 1982. Cetus was disappointed but undeterred. A potential way around a competitor's patent claims on the "natural" gene and protein sequences was to mutate inactive amino acids, creating a "mutein." Cetus was having success with this approach and by May 1983 Cetus was making IL-2 in batches one hundred thousand times larger than those Rosenberg's lab was receiving from Du Pont.

In June 1983 Cetus held its annual retreat. Rosenberg was invited to give a lecture. At that session, Cetus management apparently decided that the potential benefits of having this renowned cancer researcher conduct trials with its recombinant IL-2 outweighed the risks arising from the manner in which he would likely conduct them. Ultimately Cetus decided to give Rosenberg as much IL-2 as it could because the National Cancer Institute had the clinical facilities and a "rapid track" with the Food and Drug Administration (FDA) for intramural clinical trials. Cetus's recombinant IL-2 could stimulate the production of T cells at concentrations two hundred thousand times greater than natural IL-2. The recombinant IL-2 was far purer and more potent than the material Rosenberg was used to working with. He muses: "a half-serious computation estimated that over the next year the drops of recombinant IL-2 left in the bottom of the test tubes that we threw out were equivalent to the amount we could have gotten from killing 900 million mice."[30] He began experimenting with the recombinant product within two days of his return to the NCI.

Rosenberg sought to assess the ability of recombinant IL-2 to make T cells grow, to measure its effects in *in vitro* sensitization of T cells against foreign cells, and to use it to make more lymphokine-activated killer (LAK) cells. These are so-called cytotoxic, or "killer," T cells, which, when activated by IL-2, destroy a variety of types of tumor cells. Rosenberg's results were spectacularly successful with mice. His human results with IL-2 alone were completely negative. By 1984 he had been allowed to begin clinical trials on terminally ill patients. All seventy-five of these patients died. Further, IL-2 proved to be highly toxic. Rosenberg was discouraged but undaunted. He had been at this threshold before. Rosenberg and his boss at NCI, Vincent DeVita, had ignored conventional medical opinion by attacking childhood leukemia more aggressively than had previously been attempted. They were successful even though there had been deaths as well as major neurological complications. Rosenberg had one remaining hope, using the LAK cells in combination with recombinant IL-2. The FDA, which had been refusing to authorize the use of this combination until each component had been tested separately (even though a separate evaluation of untreated LAK cells would be therapeuti-

cally futile), finally relented and on 4 October 1984 granted Rosenberg permission to proceed with his trials. The upcoming set of
trials was either the end of the line for this strategy or a new beginning. Rosenberg pulled out all the stops: "I decided that the next
patient would be treated like the animals were."[31]

Focus.

THREE

PCR: Experimental Milieu + the Concept

The arrival of Fildes both symbolized and focused the imperative to develop commercial products at Cetus. One group that was particularly reminded of this mandate was Henry Erlich's Human Genetics group. Erlich joined Cetus in part for personal reasons (he wanted a job in the Bay Area), and in part because his scientific agenda—the interrelations of genetics and immunology—was supported and warmly encouraged by Cetus management. While maintaining his research agenda, Erlich has accommodated the company's demands for commercial products (diagnostic tests, forensic tests). He has, however, refused to move his interests in the HLA (human leukocyte antigen) system, highly polymorphic components of the immune system, into the background, or conceive of them as his own "private" work, in the way Tom White did for his evolutionary interests. Erlich observes:

> I made the decision to join Cetus largely on the basis of personal interactions; I like and respected David Gelfand . . . and Tom White. After I had made the commitment to come to Cetus and I was finishing up my research at Stanford, it was a very difficult time for me because I really didn't know if I'd made the right decision. I remember having lots of sleepless nights. How had I ended up going into the profit-making commercial arena, because I never imagined I would—and I was quite anxious. In fact, I didn't really know what Cetus was like— I didn't know what directions they were taking. I thought about the fact that it could conceivably be very exciting to

try to do biological research that led to practical, useful things. . . . Even before I started work at Cetus, I had the idea for doing something which is now called HLA DNA typing. I remember this exact moment. . . . I was meeting with Ron Cape, and I thought, "Well, if they think this idea is something I might be able to work on, well gee, maybe this would be an okay place to work. I didn't think they'd go for it because it was a totally wild idea at the time. They did. And that's really, I think, the crux of how one can ever really establish a satisfying scientific life in a company. If you find a project that is of real fundamental passionate interest to you, and if it also has the possibility for having some real practical commercial outcomes. Then you have the prospect of a company project that really satisfies a scientist.[1]

In the summer of 1982, striking out in a new direction, Cetus formed a group to develop DNA diagnostics. While part of Erlich's lab continued its work on the structure and function of the HLA genes, some resources were transferred to improving existing technical methods with the goal of producing diagnostic tests. A young technician, Randall Saiki, was assigned major responsibility in the project. In the summer of 1981 he moved to Human Genetics from Microbial Genetics, where he had been working for two years since joining the company in 1979. The HLA typing work was tedious and Saiki was eager to move on. "It was too abstract for me. It was just bands on a gel, and somehow this band was related to a type which could be related with this disease."[2] Erlich appreciated Saiki's frustration and agreed that Saiki should devote more time to technology development. In the early months, the group experimented with a number of different ideas and methods to achieve a generalizable diagnostic procedure. The goal was to reduce the "art" involved in available methods. Though such art was acceptable in research, it presented an obstacle to commercializable diagnostics kits. For example, Saiki undertook a series of experiments to see whether an oligonucleotide probe could be directly hybridized to DNA in solution and then filtered out and identified. It took close to six months to decide that the approach and its variants were not going to work sufficiently well. In retrospect, the audacious, "solve-it-all-in-one-kit" strategy had had a low probability of success: too many different steps were being consolidated too rapidly, and not enough was known about each of the many elements (enzymes, reagents, experimental conditions, etc.) involved. The "go for broke" approach, while technically dubious, was encouraged by the expectations of the media, corporate leaders, and the medical establishment for a magic bullet, and by the ready availability of funds for experimentation, equipment, and personnel.

THE SYSTEM: THE BETA-GLOBIN MUTATION

Stymied, the group members decided to reverse strategies. They agreed to move to a *known system* and convert it into a model system for genetic diagnosis. They chose as the system the beta-globin mutation (underlying sickle-cell anemia). By the early 1980s sickle-cell anemia was a well-studied case. The gene sequence coding for hemoglobin was known, and the disease-causing mutation had been identified as a single base-pair change at a known chromosomal site. For a diagnostic system, the beta-globin mutation represented the perfect challenge: a single base-pair change in a single gene in a sample of total DNA.

The beta-globin mutation site had one highly distinctive characteristic: it happened to correspond to a restriction-enzyme cleavage site. What is a restriction enzyme? Certain types of bacteria possess a powerful adaptive mechanism, the ability to protect themselves against DNA from other bacteria by producing enzymes that excise or "restrict" segments of foreign DNA at sites of specific sequences (usually four to six base pairs long). The host cell that produces the enzyme usually modified its own genome so that the enzyme does not recognize the site. Researchers therefore dubbed these proteins "restriction enzymes." Evolution had provided the bacteria with an intriguing adaptive mechanism. Humans put the mechanism to new uses, removing it entirely from its original context and turning it into a submicroscopic tool for cutting DNA into smaller, experimentally more manageable fragments.

Restriction enzymes cut DNA at intervals, yielding fragments of specific lengths called, not surprisingly, restriction fragments. A researcher named Edward Southern invented a method to identify and compare the lengths of these restriction fragments. In Southern blotting, fragments of DNA produced by restriction enzymes are placed on a gel. When an electrical current is applied to the gel, in the technique called electrophoresis, fragments migrate across the gel at differential speeds depending on their size, i.e., smaller fragments move more quickly and cover more distance in less time. The DNA is then absorbed by, or "blotted" onto, a membrane that maintains the relative positions of the DNA fragments. A radioactive probe is then poured over the membrane. The probe binds only to the bands containing sequences to which it is matched. After careful washing, X-ray film is then laid on top of the membrane, where the radioactivity exposes the film and makes an image of the fragments to which the radioactive probe has bonded. When the film is developed, after several weeks of exposure the radioactive bands appear as dark lines.[3]

The Southern blotting method rapidly became an indispensable tool in molecular biology. It is directly relevant to PCR, not only because of its general methodological importance, but because Kary Mullis hated using it. He disliked the number of steps involved, the time required to perform them, and the radioactivity required. Other scientists shared Mullis's antipathy to Southern blotting, but, given its power and the lack of alternatives, they accepted it as a part of the practice of laboratory science. Throughout this period, Mullis was continually searching for alternatives to Southern blotting.

In January 1983 a landmark paper announced a successful diagnostic test for sickle-cell anemia that used two probes to distinguish the normal beta-globin allele from the mutation. The DNA from individuals homozygous for the normal beta-globin gene hybridized with the beta-globin probe; the DNA from those homozygous for the sickle-cell beta-globin gene hybridized with the sickle beta globin–specific probe. The DNA from heterozygous individuals hybridized with both probes.[4] This was the first time an allele-specific oligonucleotide probe had been successfully used on human genomic DNA.

Although the test was a good one for beta globin, the method still contained important limitations. Success had been achieved by pushing the radioactivity of the probe to a maximum. Even then, although the highest practically possible were used, the signal produced was very faint, even after a week's exposure. The overall hybridization efficiency was low. In this instance, since the researchers knew which fragment to look for on the gel, the problem of nonspecific hybridization could be minimized. These results showed that a successful test could be achieved but also demonstrated the limitations of available methods, limitations that were particularly salient for biotech scientists operating under commercial constraints.

SENSITIVITY: OLIGOMER RESTRICTION

Erlich's group was directly interested not in commercializing the sickle-cell test but in generalizing a diagnostic method. Therefore, once the decision was made to switch to beta globin as a model system, work proceeded on two parallel tracks. The first involved the development of a nonradioactive probe to replace P^{32}, the highly radioactive phosphorous probe. Not only was it desirable to avoid radioactivity in general because of its potential health risks, but any commercial diagnostic kit ideally should contain probes that could be stored for comparatively long periods of time and that could be reused. The nonradioactive probes available at the time, however, lacked the requisite sensitivity for the beta-globin system. The second track of research was aimed at developing ways around the Southern blot. The ultimate goal was to invent a simple, general test that could be used for multiple diseases—and carried out in a single test tube.

The group succeeded in bringing these elements together into a single system—the oligomer restriction (OR) assay. The OR assay built on the model system developed for beta globin in straightforward steps, although it took a good deal of work to make the system function reliably and efficiently. The first step was to denature the sample (heating it so as to separate the double-stranded DNA) and then to add a (radioactively) labeled probe to hybridize to a single strand in the sample. If the target DNA in the sample had the normal sequence (and therefore contained the recognition site for the restriction enzyme), the DNA would be reconstituted. The second step was to cut the DNA with a restriction enzyme a second time, thereby releasing a small radioactively labeled fragment. If only the sickle cell causing mutation were present, one would see only the probe (the fragment would not be released). It was technically very simple to separate the small fragment from the large probe and to infer whether the normal gene or the mutant was present in the sample.

Although the results were promising, the system remained insufficiently sensitive. The probe hybridized not only to the beta-globin gene but to other genomic sequences as well. Therefore, a further step was required in order to separate the beta-globin signal from the other hybridization products. The hybridization was performed close to the melting temperature of the probe. Under these "stringent" conditions, probes that are not well attached to their target will peel off. Stringent conditions maximize discrimination; however, high discrimination yields low overall hybridization efficiency.

Several aspects of the system remained "empirical," i.e., unpredictable, such as the annealing temperatures of short oligonucleotides and the denaturing temperatures of DNA of different sequence compositions. There were problems with labeling probes, gels, and other experimental conditions. Verifying the proper conditions for each element could take a month or more. The work was drudgery. The core challenge remained how to increase sensitivity and reduce background noise. No one saw that there was another route to solving the "sensitivity" problem. As Saiki put it: "Everyone was trying to amplify the signal; no one was thinking about ways of amplifying the target."[5]

The Inventor: Kary B. Mullis

In 1979 a position became available in the DNA synthesis group at Cetus. The laboratory was headed by Chander Bahl, a student of the famous synthetic oligonucleotide chemist Saran Narang, a collaborator of H. Gobind Khorana. The DNA synthesis group provided a basic technical service, making customized oligonucleotides for other labs. Tom White convinced David Gelfand, recently appointed vice president of Recombinant Molecular Research at Cetus, to interview Kary Mullis, a biochemist White had known as a graduate student at Berkeley, for the position. Mullis, bored with the lab position he held in San Francisco, was eager for a more challenging situation. Although Bahl expressed reservations about Mullis's lack of experience with DNA synthesis, he acquiesced and Mullis was hired.

Kary Mullis was born 28 December 1944 in Lenoir, North Carolina, and raised in Hickory, a small town in South Carolina. His father, like his mother a native of rural South Carolina, was a salesman for the Southern Desk Company and traveled around the state selling institutional furniture, including lab equipment for high schools. Speaking of him, Mullis says:

> He's a real nice man, personable kind of guy. But he liked to travel. I like that too. It's a funny thing to think of inheriting, but I travel now, too. I like consulting. Consulting is kind of the same as being a salesman. You come in, you spend a little time with this one client, and then you leave and you come back in a month. You slowly build up a friendly relationship with them, but you don't get involved in all their pains and all their misfortunes. You help them solve their problems if you can, and then you ride off into the sunset.[6]

When Mullis was five and a half, the family moved to Columbia, the state capital, for his father's business. During this period his mother became more involved with community life, PTA, civil defense work. "She was always into all that stuff, any time there was some civic little thing like that, and she ended up knowing everybody [in] Columbia." Later, after a separation, with the expenses of college education for her sons looming, she began selling real estate, starting small and then developing "one of the biggest real estate companies in South Carolina."

Mullis enjoyed school and did well, encouraged by his family and teachers. In his junior high school years, Mullis's fondest memories are of making rockets with his brothers:

> We made a rocket out of things like pipes and stuff we could find around in the yard. We finally had a spot welder that we put fins on with. . . . We had a design that

would go up over a mile and come back down, with a little frog in a 35 millimeter film canister, all wrapped in asbestos and stuff, with a parachute to get him back alive. . . . There was an old sandlot that we could fire them in. I don't know who owned it, and nobody bothered us there. We'd set the woods on fire occasionally, but we'd always put it out.

He is grateful for the encouragement he received from one of his science teachers. "It was like 'If you want to study general relativity, you can study it. You can study it right here.' You don't have to wait until after four years of college or something. If you want to know what it's like, here's a book. See if you can figure it out. Even as a kid, I could sort of see what I would call the deep issues involved in things, like what is the meaning of arithmetic." He has fond memories of a creative writing teacher who encouraged him to give his self-expression a literary form. "So by the time I left high school, I felt like education was something that I could get myself, and I liked libraries." The natural environment was basically benign, the social environment was supportive, freedom was a personal affair. Social and political issues are not directly referred to in Mullis's self-portrayal.

Berkeley

In 1962 Mullis followed in his older brother's footsteps, enrolling at Georgia Tech and majoring in chemical engineering. He did well in school, cultivating a strong interest in physics (including cosmology) in addition to chemistry. When a friend favorably described Berkeley to him, he decided to apply to graduate school there. Mullis was surprised to be accepted at such a prestigious institution, not having any personal connections or, as he puts it, "people to call up." Berkeley was "like being let out of jail." There was no one in Georgia "synthesizing new molecular forms of psychedelics and saying, 'What do these things do? How does it feel if you take this stuff and put in it your mouth?' In all the chemistry labs and biochemistry labs at Berkeley there was a little element of that going on; like when I first arrived there, they hadn't decided that these things were bad for you and that they were going to be illegal."

In 1966 Mullis entered the doctoral program in biochemistry at the University of California at Berkeley. Initially his advisor was Allan Wilson, who allowed him to pursue a wide-open course of study. Mullis took very few molecular biology or biochemistry courses, figuring he could pick it up from his friends. Instead he explored widely in the sciences. Eventually he began spending more time in the lab of J. P. Neilands. Mullis thrived in the milieu of Neilands's tolerant egalitarianism. Although Neilands was an antiwar activist, Mullis viewed Neilands's commitments in moral, not political, terms: "His convictions are always really moral, upright kind of things. He's a fine person." These character traits were combined with an ambiance of freewheeling discussion; Neilands's was not a high-pressure, single-minded, careerist lab. "Every afternoon about four Joe would come in the lab and he'd have a big four-liter beaker full of water and make tea, and we'd go in the tea room and we'd talk about whatever Joe felt like or somebody felt like talking about." Mullis began spending more and more time at the lab. Neilands imposed minimal formal requirements on his graduate students. He expected rigorous scientific work but didn't expect them to be maniacally committed to advancing either his own research or their own. Mullis took full advantage of this environment:

> I had been, like everybody else there, taking LSD all the time, and it was always a stimulating kind of thing intellectually. You'd come down from an LSD trip with all kinds of ideas, just scrambled all up, and you'd try to pull them out. At one point, I started thinking of a cosmological theory that I felt like was a little more solid than the Big Bang or steady state. In fact, in my eyes those two were kinds of viewpoints that you might have if you tried to look at the universe from the outside, but that was not a legitimate real way to look at it, and that's why you ended up with two paradoxical things that you couldn't decide between. I conceived of a way that you could, if you look at it from inside, you'll see that those elements of it, that it will look that way, but it's not really either exploding at a moment or forever expanding. Those aren't really good ways to talk about it because you're in

it. If you put that element back in it, you come up with this little theory.

Mullis submitted an article entitled "The Cosmological Significance of Time Reversal" to *Nature* in May 1968. The article was eventually accepted after two rejections and a lively exchange of correspondence with the editors. "It was unprecedented, sort of, for a graduate student to publish in *Nature*. I did it because I didn't realize it was so unprecedented. I thought, you have an idea, you send it off to *Nature*." Mullis attributes his success in his doctoral preliminary exams, for which he admits he was not thoroughly prepared, to the *Nature* article, which he feels made it impossible for his committee to fail him.

In 1972 Mullis received a Ph.D. in biochemistry for his thesis "Structure and Organic Synthesis of Microbial Iron Transport Agents." The research entailed a good deal of experimentation with organic synthesis. "I enjoyed the idea of learning how to make something that had never been made before. . . . I didn't care about that substance, particularly; it was just one thing. But it was just the right size and the right kind of molecule, and I learned a lot of chemistry." The tone of the thesis was, as Mullis puts it, "a narrative light humor," which occasioned a certain amount of debate (strong dissatisfaction from some, outright criticism from others) among the members of his dissertation committee, but Neilands's strong support carried the day.

Shortly after, Mullis accompanied his new wife to Kansas, where she entered medical school. He saw no harm in postponing his scientific career, thinking he would write a novel. "I tried and discovered that I didn't have enough experiences in my life to write a decent novel. I couldn't figure out how anybody could ever be unhappy. It was not going to be a decent novel." Mullis accepted employment in a pediatric cardiology laboratory, where he assisted in work on the possible biochemical basis of a chronic lung disorder in children. This experience was a characteristic one for Mullis in several respects: the priority of emotional attachments over career considerations; his willingness to accept a position of lower status than a Berkeley Ph.D. might otherwise expect; his enthusiasm for expanding both his scientific and technical capabilities, in this instance in physiology and medicine. Equally typically, Mullis's learning curve rose rapidly and then leveled out, leaving him in a state of frustrated restlessness. As Mullis puts it, he became fed up with slaughtering animals, ending each day with "a bag full of rats' heads and other gross things." He began writing science fiction. In 1975, after separating from his wife, Mullis returned to Berkeley with his future third wife, a nursing student from Kansas. He worked for close to two years as the manager of a local restaurant and coffee shop, The Buttercup Bakery, owned by his first wife. In 1977, at the Buttercup, he crossed paths again with Tom White. White passed on to Mullis the news of a job in the laboratory of Wolfgang Sadee, a specialist in pharmaceutical chemistry, at the University of California at San Francisco Medical School. The position was for a chemist interested in isolating endorphins, opiate-like molecules produced in the brain. Again, Mullis accepted a situation parallel to the one he had left in Kansas, insofar as it offered initially challenging work and a status lower than his credentials warranted. Mullis enthusiastically availed himself of the opportunity "to bring himself up to speed" in current research on the brain and drugs. Repetition—Mullis soon found himself slaughtering laboratory rats.

Cetus Corporation, 1979–81

When White told him about an opening at Cetus, he applied. For Mullis this was the period that "DNA synthesis was starting to be interesting because finally DNA was becoming chemical." Synthetic work and cloning were coming into the picture. "The stuff is alive now. We've made an organic chemical piece of DNA. It was significant, I thought. So I went to the literature and learned how to make those—I mean, I figured I could learn how to do it, how to synthesize oligonucleotides—and I did."[7] Again Mullis's learning curve ascended rapidly, and he was soon bored by the laborious and repetitive operations required to make the oligonucleotides. Mullis set himself the task of improving the speed, power, and efficiency of the available methods. His irreverence, verging on belligerence, and his increasingly trenchant criticisms of current procedures produced clashes with Bahl. However, Mullis was able to demonstrate the superiority of his approach by

producing larger quantities of the oligos more rapidly and efficiently. He recalls:

> Probably I worked harder in that time in my life than I
> ever had in terms of hours. I was very interested in the
> job. It was really fun to learn how to synthesize DNA. It
> was just organic synthesis, pure and simple. No rats. And
> it was an exciting time at Cetus.... They had just gotten
> enough money to really start hiring a lot of people and
> doing a lot of things. It was the heyday of biotechnology.
> There were all kinds of bold ideas floating around all the
> time about what we were going to make, and there was
> absolutely no restraint in terms of imagination.... The
> company was really fun.... They were right in the mid-
> dle of something that was a red-hot kind of an area. It got
> worse after about three years. It got more into the business
> kind of stuff, but it was really fun at first.[8]

In April 1981 Tom White was appointed head of the Recombinant Molecular Research department. He appointed Mullis as
head of the DNA synthesis lab, replacing Bahl, whose work White
found too plodding. Mullis's mandate was to speed up and improve oligonucleotide production. Soon there were a series of con-
frontations with scientists and technicians in other laboratories
who complained about the quality and consistency of the oligos
Mullis's lab was supplying them. Mullis had decided to stop sequencing every oligonucleotide he delivered because sequencing
was still a very time-consuming as well as highly radioactive task.
In order to sidestep this drudgery, Mullis had devised a mathemat-
ical algorithm that could use an ultraviolet absorbency measure-
ment to calculate what the expected absorbency of a particular sequence composition should be. Although innovative, the method
was untested, and other Cetus scientists were reluctant to rely on
it. Scientists encountering problems in their cloning experiments
attributed the cause of the problem to the quality of the oligonucleotides provided by Mullis's lab. Mullis attributed their problems
to their abilities.

He firmly believed he could improve the speed of operations
without sacrificing quality. He trusted his intuitions; he expressed
contempt for the kind of proof other scientists expected. The fact

that Mullis had not run a standard series of controlled experiments only served to confirm the doubts and fuel the animosities of his detractors. Push came to shove. White ordered him to run controlled experiments. He did, and they worked, vindicating Mullis's assessment of his own talents. His success reinforced White's view that here was an innovative scientist. White recalls that Mullis had proposed

> at almost every scientific retreat a number of wild ideas, some of which were flatly wrong because he wasn't really familiar with some of the most basic aspects of molecular biology. And also because he was abrasive and combative and often times his comments would be counterproductive in meetings where people have to try and work together. Mullis had a grudge against his critics and they had a grudge against him.[9]

Internal tension in the synthetic oligonucleotide lab intensified when a relationship between Mullis and another chemist working in his lab turned openly and publicly conflict-ridden, its *sturm und drang* played out amidst the lab's day-to-day work. In one incident, Mullis became jealous of what he interpreted as amorous intentions on the part of a technician in another lab toward his woman friend; at one point, she phoned White to say that Mullis threatened to appear at Cetus armed with a gun. White finally intervened, not to fire Mullis but to assign a mandatory week's vacation to the endangered party. There were other such incidents.

Mullis proceeded to explore means of improving the synthesis of oligonucleotides. As he puts it, "I like to play around in the lab, and so I had all these oligonucleotides, and I really didn't know too much about the properties of oligonucleotides because nobody really did."[10] He tinkered with the variables involved in the synthetic process. The denaturation and renaturation qualities of DNA not only were little understood, they were quite irregular. Mullis first explored temperature and time. Playing with a series of computer programs, he attempted to predict melting temperatures for given sequences. If these thresholds could be established, it would be logical to follow the same procedures for renaturation temperatures. At the time, Mullis was working with a spectrophotometer that could distinguish between a bonded pair of bases and a denatured set of bases. The machine had a temperature-control device on it so that as the temperature rose, Mullis could see exactly the point at which the bonded pairs separated. It became apparent during these experiments that the temperature could be varied rapidly without dramatically altering the results. "I came away with the idea that oligonucleotides hybridize fairly rapidly."[11] It became clear that the time allowed for hybridization could be dramatically reduced.

Another important technological innovation taking place during the late 1970s and early 1980s was the introduction and improvement of DNA synthesizing machines. Mullis credits a company called Biosearch for showing him the first prototype. A friend, Ron Cook, a peptide chemist from UCSF, brought one to Cetus to show Mullis.[12] Although the machine was not fully successful, it was still enormously timesaving, reducing a month's to a day's work. Mullis was active in suggesting improvements on the prototypes, especially in writing some of the software. The result was that Mullis's lab was one of the first in biotechnology to move toward the automation of DNA synthesis. Once the machines were introduced, productivity jumped by a factor of ten and continued to rise. This increase of efficiency gave Mullis added time to do more computer work; he devised programs to simplify the administrative tasks of the lab:

> I had just started computerizing everything in the lab. That was what I did for fun, and also it did all the work, and it was like I didn't have to do anything. After a while, I could watch the lab from my terminal at home, and I could see how things were doing and I could find out all kinds of information from just having all the analytical instruments in the lab tied to computer routines that people would have to run that would make files and I could look up and find out what had happened—and nothing much exciting was happening anyhow.[13]

Mullis became intrigued by iterative processes known as "loops," which were well known to computer programmers (computers were designed to perform repetitive tasks rapidly) but foreign to biochemists, who rarely thought of repeating an operation over and over again. Tinkering with iteration, both at work on the powerful VAX and at home on his Amiga, Mullis began to explore exponential amplification—perhaps the key original element in his overall conception of PCR.[14] As Corey Levenson, one of Mullis's lab mates and best friends, puts it:

> Kary was playing around a lot with fractals at that time. He was working on a piece of paper with these patterns which were being generated by a very simple mathematical equation, just each time the results were put back in again and fed through. Maybe the concept of some discrete event occurring on the small scale and then being translated onto the larger scale prepared the way.[15]

PCR YEAR ONE: SPRING 1983–JUNE 1984

Mullis's own version of how he came to the concept of PCR is presented in several places: in an article he published in the April 1990 *Scientific American,* in his sworn testimony at the patent trial between Du Pont and Cetus during the winter of 1990–91, and in subsequent interviews and publications. The proverbial moment of discovery came on a Friday night in the spring of 1983, during the several hours' drive on winding state highways and rutted country roads from Cetus to Mullis's cabin in Mendecino County.[16] Mullis can only date the fateful drive to either the spring or fall of 1983. He is certain, however, that "the buckeyes were in bloom." The *Sunset Western Garden Book* describes buckeye: "*Aesculus:* Deciduous trees or large shrubs. Leaves are divided fan wise into large toothed leaflets. Flowers in long, dense, showy clusters at the ends of branches. *A. californica.* California Buckeye. Native to dry slopes and canyons below 4,000 ft elevation in Coast Ranges and Sierra Nevada foothills. Striking sight in April or May when fragrant, creamy flower plumes make it a giant candelabrum."[17]

During his drive, Kary Mullis was thinking conceptually, soaring above the daily grind of making experiments work, escaping the frustrations of the lab, where the gap between elegant concepts and successful experimental systems was every scientist's demon. Mullis was thinking about the stubbornly challenging issue of sensitivity in the beta-globin project. What would it take to arrive at a general procedure for identifying a single nucleotide at a given position in a DNA molecule?[18] His train of thought brought together DNA polymerases and DNA sequencing.

During the 1950s, scientists had identified the polymerases, the class of enzymes required for the repair and replication of DNA.[19] However diverse and complex their operation in nature, the principles of polymerization are straightforward. The starting point of DNA replication is separation of the double-stranded DNA into single strands. This separation operates as a normal part of the cell-division process and can be mimicked in the laboratory through heating. Once the double-stranded DNA is separated, the duplication process can begin. Several constraints operate on this process. Polymerases cannot begin their duplication work on a single strand of DNA alone; they require an attached anchor. This anchor, this extra strand of nucleic acid, is referred to as the *primer*. The single strand to which it is attached is the *template*. The polymerase builds upon the primer along the template. In order to accomplish this extension, the polymerase requires building blocks, called nucleotides. These building blocks are available in the cell, but under laboratory conditions they must be added. The polymerase performs this primer extension in accord with a simple principle called *complementarity*. DNA is a double-stranded helix held together by hydrogen bonds between specific pairs of bases. The four bases (adenine, thymine, guanine, cytosine) bond with each other in a fixed complementarity (A-T, G-C). One further principle of DNA replication completes the picture. The double strands of the DNA molecule are *antiparallel*. Although each strand of DNA is composed of identical bases, bonding takes place only in one direction. By convention, the three-prime end of one strand pairs with the five-prime end of the other strand and vice versa. This antiparallelism is important because it provides the route along which the polymerase proceeds and the point at which it terminates its operation.

Mullis was thinking about what would be necessary in order to arrive at quicker, cleaner, more efficient ways to identify the single base-pair mutation in the beta-globin gene that causes sickle-cell anemia. The general diagnostics goal was to identify a specific sequence in a high-complexity target with only small quantities of the target. In order to carry out such a procedure, one needed a reliable means of isolating the target region from the complex and delicate DNA molecule, and then a means of identifying which allele, or variant, was present in the isolated target. Mullis and others had been thinking about employing modifications of the then standard DNA-sequencing technology with the goal of increasing its sensitivity and efficiency. Sequencing methods identify single base pairs, hence their relevance. Attempting to solve the "sensitivity issue" in a general manner, Mullis was performing thought experiments on a series of known techniques, principles, and strategies. Sequencing proceeds from, and capitalizes on, the principle of base-pair complementarity. The basic idea is that if one labels one side of the complementary pair, then the base to which it bonds will be known as well. There were a variety of ways to label the oligonucleotide probes used to hybridize with, or attach to, the target. Mullis's lab provided such probes to other scientists at Cetus.

Mullis was musing about an idea that had been bandied about his lab: what if he used two oligonucleotide primers instead of one to bracket the target base pair? If these primers could be made to attach to each of the antiparallel strands, and if each of the oligos were a different size, then the procedure would contain mutually confirming sequencing information from both strands. Mullis writes, "Although I did not realize it at that moment, with the two oligonucleotides poised in my mind, their three prime-ends pointing at each other on opposite strands of the gene target, I was on the edge of discovering the polymerase chain reaction."[20] He was only at the edge, and not over the edge, because he was still thinking about the diagnostics problem: how to increase sensitivity?

Wondering about potential difficulties led Mullis to take the next big step in his thought experiment. Since stray building blocks are frequently found in solution and could well be misincorporated by the polymerase, thereby producing false information about the target, it would be helpful to find a way of destroying them. Mullis wondered what would happen if he were to employ the polymerase twice. By adding the polymerase the first time he would eliminate or reduce the extraneous building blocks from the sample because they would be incorporated by the polymerase into an extended oligonucleotide. By heating the solution, these extended oligos could then be separated from the target DNA. Although these extended oligos would remain in the solution, Mullis reasoned that there would be fewer of them than of the target DNA. It seemed plausible that the targets would still hybridize with the unextended primers, supplied in great quantity, when the mixture cooled. Mullis was, in a sense, reversing the typical sequencing reaction—using up or incorporating the building blocks instead of adding them.

Mullis hoped that after this preliminary cleanup the primers would go back to the designated spot on the template and be in the correct position to receive a labeled building block. Then one could add specially designed and labeled building blocks and more polymerase. If the polymerase extended the primer down to the point where the original primer had been anchored, as it was supposed to do, he would have two copies of the original template. "Yet," he recalls,

> some nagging questions still nagged at me. Would the oligonucleotides extended by the mock reaction interfere with subsequent reactions? What if they had been extended by many bases, instead of just one or two? What if they had been extended enough to create a sequence that included a binding site for the other primer molecule? Surely that would cause trouble—I was suddenly jolted by a realization: the strands of DNA in the target, and the extended oligonucleotides, would have the same base sequences. In effect, the mock reaction would have doubled the number of DNA targets in the sample![21]

In Mullis's after-the-fact reconstruction of his thought experiment, this was the breakthrough moment. His tinkering with fractals and other computer programs had habituated him to the idea of iterative processes. This looping, back and back again, as boring and time-consuming as it might be on the level of physical practice, was nearly effortless on the computer. Mullis made the connection between the two realms and saw that the doubling process was a huge advantage because it was exponential.

One last step remained: *specificity*. In the first extension reaction there was nothing to tell the polymerase to stop extending the primer. Mullis reasoned that in the next and all subsequent rounds of his chain reaction, there *would* be something to stop the extension process. The so-called long products could only be extended so far—to the place that the other primer began. After that point there was nothing to extend because there was no more template. The reaction would produce products of defined length, exactly the length between the outside ends of the two primers. The reaction exponentially produced a defined and specific target. "Eureka! By then, I'd forgotten all about the problem of diagnosis. I realized that this is a process that I can use anywhere I want to amplify a little section of DNA and I can take the oligos and I can move them just by changing their sequence."[22]

After a weekend of scribbling on all the available paper in his cabin, Mullis returned to Cetus. Fired up about his idea, he remained just a little cautious, wondering why, if the procedure was so simple and so powerful, no one else had thought of it. His first step, therefore, was to find out if anyone had. A literature search by the Cetus librarian produced nothing. Mullis began asking scientists and technicians both within and outside Cetus whether they knew of any existing technology capable of achieving similar results. No one did. From the start, Mullis was publicly loquacious, exhilarated, exuberant about his idea. Any number of other people both within Cetus and without could have followed his lead, scooped him. No one did. No light bulbs went on.

Heartened by the negative finding about prior art, Mullis began to do more systematic reading about polymerases. Although convinced he was heading in an original direction, Mullis found very little in the technical literature that would help him to establish the kind of experimental parameters he would need. He spent the summer of 1983 imagining things he would like to know but not actually doing any experiments that could help him find out. Exactly why Mullis did no experiments for three to five months after his Eureka! experience remains unclear. Several factors may have contributed to his inaction: like everyone else at Cetus, he was busy; emotional strain and upheaval in his love life spilled over into his professional life; and the response to his idea from his colleagues at Cetus was nearly unanimously discouraging. His descriptions simply did not arouse his friends' and colleagues' curiosity.

In August 1983 Mullis first presented the concept of PCR at a regular Cetus seminar.[23] The initial response was negative. A few lab technicians expressed interest. Mullis remembers, "Most of the people either left the room before I was done, or—sort of stayed around to heckle. You know, they were friends of mine. They were used to me having really bizarre ideas, and they just thought this was one more of those."[24] The fact is that Mullis's credibility was low at Cetus. This was not the first revolutionary idea Mullis had convinced himself he had discovered. In the notebook in which he recorded the first attempts at experimentally demonstrating PCR, there are other experiments with even grander horizons: "On page 141, I'm curing cancer."[25] Because Mullis had no experimental data whatsoever for the feasibility of his concept, not many people were willing to give him the benefit of the doubt. Those, like White, who were most sympathetic to Mullis were working extremely long hours during this period, engaged in the grueling races to get the interferons and IL-2 into clinical trials. Seventy to ninety-hour work weeks were not uncommon. Fildes's pressure for commercialization only intensified an already hectic schedule.

During the summer of 1983, or even later, during the more intense experimental phase of 1984, no one ever raised the issue of prior art. For Cetus scientists, technicians, and the distinguished board of scientific advisors, Mullis's idea was unequivocally unexplored territory. Corey Levenson, who worked in Mullis's lab in 1983, remembers:

> When Mullis first presented it to people on paper everyone was looking at it and trying to figure out why it wouldn't work. It was so simple it didn't take a lot of explanation. Once they understood it they thought there must be some reason that it didn't work. The attitude that everyone took was "it looks good on paper but show me a band on a gel. Let's see you reduce it to practice." Everyone had this feeling that there is something that they are not quite smart enough to see, but there is a reason why

this wouldn't work. Including Kary. Why hasn't someone
else thought of this?[26]

The Concept + Available Techniques

From September through December 1983, Mullis performed a series of experiments that started from his concept of PCR, and
sought to make PCR work using available techniques. During this
first period, Mullis introduced no innovation on the technical
front; the reagents were either commercially available or in general use at Cetus, or widely known practices in the trade. Mullis
gave little if any preliminary attention to his choice of experimental conditions or system. It was only at the end of this first three-
month period, when the experimental results were clearly unsatisfactory, that he began to give more methodological attention to the
model system.

On 8 September 1983 Mullis performed his first laboratory experiment, recorded as "PCR01" in lab notebook number 1,000. He
went for the jackpot, attempting an improbable, all-or-nothing approach. He chose the human nerve growth factor gene as his target
because scientists at Genentech had just published its sequence.
Objectively, the prospects of success were low; not only had he
chosen to work with a very complex human genomic DNA, but
it was a single-copy gene, a very difficult target. Mullis's plan was
to isolate a section of the gene approximately 400 base pairs long,
attach primers, and extend them. He put the two oligos and the
four deoxynucleoside triphosphates together in a test tube. He
chose a commercially available buffer that had been used for polymerase work and modified it according to a formula in use at
Cetus. He heated the solution, cooled it, and went home, hoping
that there would be extension, separation of the fragments, rehy-
bridization, and more extension. Returning twelve hours later,
Mullis heated the solution again and added radioactive dCTP,
which he hoped would initiate a further cycle, be incorporated,
and allow the synthesized product to appear on an autoradiograph.[27]

Mullis began the experiment at midnight:

> [i]t was a shot. . . . I thought that if everything works out
> exactly right then the easiest thing I will have to do . . . is

> put those things in there, heat it up, cool it off . . . and go
> home. That's what I did. . . . I was hoping that some time
> during the night the old nucleotides would land on their
> target and be extended and the products would come off
> and the other nucleotides would come back. All the same
> process would take place, but without me having to shift
> the temperature or take anything out. It was a possi-
> bility."[28]

But there was no jackpot, there was only a great smear on the
gel. Experiment PCR01 neither proved nor disproved anything
about PCR.

In the experiment labeled "PCR02," performed in October,
Mullis retained the same reagents and procedures. However, instead of a one-step procedure, he proposed five (or ten) cycles of
heating the solution, cooling, and adding the enzyme.[29] Only at
that point would he add the radioactive dCTP and repeat the cycle
one more time in order to allow the genetic material to incorporate
it. Taking into account the complexity of the genomic DNA,
Mullis performed purification techniques and added an additional
"digestion" step. He added a further step to "clean up" other elements in the solution that might be hindering the amplification he
sought. Mullis's entry in his lab notebook is laconic: "Achieved a
negative amplification."[30]

These experimental failures in no way invalidated the concept
of PCR. Like any laboratory scientist, Mullis drew a clear distinction between the concept and what he called "analytic procedures,"
experimental protocols. He writes:

> I assumed from the beginning that this mechanism, the
> polymerase chain reaction, could work under some condi-
> tions. I didn't know what conditions those would be, and
> whether these were them I don't know for sure. But I
> was working on both the analytical method and thinking
> about the reaction itself. And until I got some result that
> I could then make better or less by manipulating the vari-
> ables, I was basically working in the dark.[31]

He continued exploring sporadically over the next two months,
concentrating on some of his older concerns, for example, finding

a way to avoid using Southern blots. As he says, "I was always interested in anything that would be a new method of analysis. I did experiments throughout this part of my career very impulsively and very rapidly."[32] In December Mullis tried using different restriction enzymes. He sought to establish measurable standards for the stability of triphosphates under heating. He began figuring out a way to establish quantitative guidelines to measure how much amplification and how much radioactive labeling would be needed in order to produce clear results on a gel.

Simplify the System

Recognizing that he was not getting anywhere with human genomic DNA, Mullis decided to simplify the system. "I said, you know, Mullis, . . . Why don't you try a DNA that's much less complicated than human DNA and try to make a smaller piece and see if you can do that. If you can do that you can work your way back up to the 400 base pair piece. The 400 base pair piece would be very, very impressive to my skeptical colleagues. That's one of the reasons I wanted to do that. I realized that was really pushing my luck."[33] He switched targets to a well-known, readily available cloning vector, the pBR322 plasmid. A plasmid is an "extra chromosomal genetic element found in a variety of bacterial species. . . . Plasmids are double-stranded, closed [circular] DNA molecules."[34] Working with bacterial DNA was considerably simpler than working with human DNA. Mullis chose a much smaller target, a 25-base-pair stretch bounded by two oligonucleotides (11 and 13 bases long). In addition to choosing simpler genetic material and a smaller target, he tried smaller volumes and reactions that were considerably more concentrated. He lowered the temperature at the next cycle to 32°C. He added less enzyme. He ran the reaction through ten cycles. He ran the results on a less sensitive gel than the radioactive tracer gels he had been using. At the end, although Mullis believes he saw a 25-base-pair band on the gel, he admits it was not distinct: "And my conclusion . . . is that there was a band there, but it's not one that would stand up to a lot of criticisms. I would definitely have to make it a little brighter."[35]

Everyone else agreed.

In order to achieve greater sensitivity, Mullis tried the experiment again with several more heating and cooling cycles and a radioactive tracer (dCTP). "This time it worked. You had a hard time seeing the band that I expected, although [in the notebook] it does say 'arrow marks the spot.' "[36] Excited, Mullis took the only person who happened to be around Cetus, patent attorney Al Halluin, into the darkroom to show him the autoradiograph. "He looked at it. . . . there were no controls; very sloppy experiment. But it was there. . . . he congratulated me, I was impressed, because no one had congratulated me on having the idea."[37] Mullis then went over to Fred Faloona's house. Faloona, with a high school diploma, had replaced Mullis's daughter, who had performed various housekeeping tasks in the DNA synthesis group, when she went off to college. Intelligent, a quick learner, and technically adept, Faloona was a part of the entourage frequenting Mullis's property in Mendecino County: in sum, a sidekick, trusted technician, and supportive collaborator. That night, Mullis's enthusiasm was oracular.

After a week's vacation over Christmas in Hawaii with White, Mullis repeated the experiment. He didn't have a 25-base-pair standard readily available, so he didn't include a control in the follow-up experiments. The gel did seem to indicate that something had been amplified even if it was impossible to say what or how much. Mullis remembers:

> I was still in a stage where this was an indication it was working but certainly did not prove it. I mean, that was exciting to me. . . . [Al] had said, you're going to have to get to work now doing some experiments that really prove that this works and we'll write a patent.[38] . . . I had absolutely no controls in this experiment. . . . If this were the last experiment I had ever done I was coming in here telling you I had done that and therefore I had done PCR, therefore, I would feel kind of foolish.[39]

Encouraged by his results, Mullis became bolder and decided to move from the pBR322 plasmid with its 4,300 base pairs to a lambda phage (a bacterial virus) system with about 50,000 base pairs. He unsuccessfully attempted a 2,000-base-pair digestion using another restriction enzyme. He tried varying times and temperatures, he made efforts to increase the level of radioactivity used in the labeling step, and he attempted various means of getting rid of background triphosphates so that any small PCR fragments would be more visible.[40]

At this stage, Fred Faloona and perhaps one or two other technicians were convinced that Mullis was pursuing something truly important or were at least sympathetic to his attempts. Although Mullis had accumulated a certain amount of experimental data as well as a certain amount of hands-on lab experience in manipulating the experimental parameters, at this point he still had no demonstrable experimental evidence that would stand up to normal scientific standards that PCR worked. Some kind of reaction, however, clearly was taking place.

Early in January 1984 Mullis shifted his target once again. Perhaps, he reasoned, custom designing the target would facilitate amplification.[41] He shifted from the lambda phage system to a 100-base-pair synthetic oligonucleotide made in his lab. During the course of the next several months, he began working on distinguishing beta-globin alleles. The results were disappointing.

In the spring—approximately one year after Mullis conceived of PCR—he returned to the 58-base-pair region of the human beta-globin gene containing the sickle-cell mutation that Erlich's group had been working on. By June 1984 Mullis was satisfied that he had succeeded in achieving some amplification. In a report he submitted at the beginning of June, covering his work from June '83 to June '84, Mullis wrote:

> This technique, involving the simultaneous extension of two oligonucleotides with a DNA polymerizing enzyme and a suitable template, will allow for the unlimited production of any given sequence, ssDNA, dsDNA, or RNA. It is presently being patented and will likely find use in diagnostics, cloning, and DNA synthesis. The technique was originally worked out amplifying a 25-bp sequence of pBR322. It has now been extended to amplifying different regions of the beta globin gene starting with whole human DNA. The number of parameters, including choice of en-

zyme, that need to be optimized is large and the work has been moving slowly due to other responsibilities.[42]

These claims would eventually prove to be true.

Normalizing Personnel: Cetus Scientific Meeting; June 1984
In June 1984 Cetus held its annual scientific meeting at Monterey, California. The meeting was an occasion for Cetus scientists to present their work as well as to brainstorm with each other and with the company's consultants. In the weeks leading up to the conference, Mullis was in bad emotional shape. His relationship with a colleague had ended. The final, passionate lovers' disputes were taking place all too commonly in public before an increasingly uncomfortable and unappreciative Cetus audience. Mullis's credibility was declining faster. At the meeting, in addition to the presentation of major projects—mostly pharmaceutical—there were fifty to sixty poster presentations. Mullis had prepared a poster showing his lab's gains in productivity. The poster also presented PCR—showing the amplification of a 58-base-pair piece of the beta-globin gene. The poster was generally ignored. Mullis remembers that the only encouraging reaction came from Joshua Lederberg.[43]

During the meeting, David Gelfand hosted a "blue margarita party" in his room. Mullis and a scientist named Mike McGrogan got into an increasingly surly match of wits, hurling accusations of scientific and technical incompetence back and forth. The escalating confrontation brought Gelfand's party to an end. Mullis wasn't done, though. He pursued McGrogan to the balcony outside McGrogan's room, where a shoving and shouting match ensued. At one in the morning, John Sninsky, who had recently come to Cetus, found himself physically mediating between the two scientists. Reluctantly, Mullis returned to his room. Still riled, Mullis called Tom White several times to tell him in an abusive tone that White had been a "jerk" for insisting that Mullis experimentally demonstrate that PCR worked when he knew perfectly well that it did. Finally, at three in the morning, White called hotel security, who escorted Mullis to the beach for a long walk. Mullis had always liked the beach.

INTERVIEW: ELLEN DANIELL

Shortly before the Monterey retreat, Ellen Daniell joined Cetus. After being the first woman in the molecular biology department at Berkeley, she was denied tenure. Daniell became Cetus's director of Personnel in 1985, and later, in 1988, the director of Business Development for the newly founded PCR division.

ELLEN DANIELL   I never felt that I was denied a position because I was a woman, but I did feel that a lot of my ignorance about how to succeed came from being a woman, being unmentored, from just not understanding what I needed to do. At the same time, I had a consciousness that a lot of what I didn't do to get tenure, I would not have wanted to do even if I had understood that it was what I had to do.

I had to make a decision to look for a faculty position somewhere else; or find something else to do that would use my science but that wasn't a university position. I was married to David Gelfand by that time; he was happy with his position. So, I never contemplated leaving the Bay Area. NSF allowed me to convert the last year of grant that I had into a salary for me to work elsewhere. Meanwhile, I had done a three-month sabbatical, beginning to get into plant molecular biology. In the '83-'84 year, with the intent of getting into a new field, I went into a lab in the Berkeley genetics department and worked on plant molecular biology. So, I had a year to make up my mind as to what I wanted to do next. I had an opportunity to apply for a grant and be co-PI, but I found I no longer wanted to plan science five years ahead. I wanted to look for the mysterious "something else" that involves science but isn't any of these other things I'd already done.

And at that point, a job was posted at Cetus for a senior scientific recruiter. I first heard about it from Tom White. I had known Cetus people for several years and I knew that the reason for creating the job was that the personnel department was considered to be completely useless at helping people recruit scientists. The R&D people wanted some real help from Personnel in recruiting. The job would lead into project management in

R&D. I was hired June of '84, in time to go to the scientific retreat in Monterey.

I was a little nervous about going in. David had been one of
the first one hundred employees. He was a big gun, well known
at Cetus. I was nervous about being defined as David's wife. But
neither in that first year or ever did I get a question about appro-
priateness. I never felt people thought I got the job other than on
my own merit.

I came in as scientific recruiter but I started almost immediately handling employee relations as well, because I understood
the science and employees felt I understood their issues. At the
end of the year I was promoted to the new title of Human Resources Manager and almost at the same day that I got the promotion, my boss resigned to go to a new company. I know of at
least one scientist who sent a message to Fildes saying, "Ellen
Daniell is a great acting director. Why don't you just make her
personnel director?" Tom was very instrumental in my getting
the job. A year and a few months after taking the job of senior
scientific recruiter, I became personnel director. Personnel had
been a second-class area. The puzzle was to make Personnel fit
the kind of company Cetus was. That's why I did it.

PAUL RABINOW   What was the state of personnel relations?

ED   I think women were well treated and for the most part
well respected. People like Price, White, Gelfand, and Erlich—
the top guys in R&D—they had grown up in an era when one
did have respect for women and equality and there were substantial numbers of women in the coed colleges they went to.

Of course, there were exceptions. An example is one woman
who was removed at her request from the lab she was in, after
feeling persecuted by people who told jokes offensive to her. She
had a hard time complaining to Personnel, and once she did, she
felt she really couldn't stay in her lab, because we needed to
speak to the people who were offending her. So, it was a negotiated chance for her to get out and for them to learn how their behavior could affect someone's work. She did well in her new position. It didn't happen again. While I would not want to say there
was no sexism at Cetus, I would say it was neither prevalent nor
condoned. The culture was much less sexist than the university.

With regard to minorities, there were very few but very capable Afro-Americans among Cetus R&D staff. There weren't many available candidates in biology at the time. There were, in contrast, a larger number of Asians in the field and at Cetus.

At that time, the Equal Opportunities Commission had mandated the need for affirmative-action plans, and we worked on a plan, which applied to women and all minorities. You had to look at the population of available different categories; look at technicians at a local level; look at Ph.D. scientists at a national level. You get all those figures from various available sources and see where you're lacking and where you need to do something.

The thing that we were able to honestly recruit people with, that turned out to be what our scientists were most excited about, the thing that they could pass on to the new people, was the fact that science was so much more cooperative than at most universities. There was also more promotion and evaluation of people on their merits; demonstrated achievement was more important than paper credentials. At Cetus someone with a B.A. could get promoted to a position of responsibility. That was absolutely impossible at the university. I think that the policy was established by Price and White, but Fildes did not disagree with it.

Personnel Threshold

In the weeks following the Monterey events, Cetus's senior scientists debated, not the potential of PCR, but whether to fire Mullis. The ball rested most directly in the court of White and Jeff Price. White's three criteria for evaluating a scientist's work were creativity, productivity, and the ability to work effectively as a member of a multidisciplinary project team. To survive, a scientist had to perform well in at least two of the three. White explains:

> So here is Mullis—creative, but none of his ideas up to that point had really been more creative than anybody else's.... And Mullis is creating havoc—affairs with people in his lab, threatening people who were going out with his friend, threatening to kill them, fistfights, threatening the guards in the evening because he didn't have his badge when he came in the building, problem after

problem. And what to show for it except wild ideas that were out of his field that people felt wouldn't work.[44]

Price and White held a meeting with Fildes to decide Mullis's fate. Their options were firing Mullis outright, removing him from the DNA synthesis group and reassigning him, or giving him a chance to test his idea. Price and White were under pressure from other scientists to fire Mullis. There was strong sentiment for not tolerating his behavior or rewarding him in any way. White decided that despite Mullis's behavior, PCR had enough potential to explore its merits further. He relieved Mullis of his responsibilities as head of the DNA synthesis group, giving him a probationary period to work on PCR. Price trusted White's judgment and agreed; Fildes supported the decision. Mullis was offered a year to work on PCR, with a mandatory progress report due in six months. Neither Mullis nor other Cetus scientists appreciated the decision. White might well have fired Mullis. If Mullis had been fired, Cetus might have continued to work on PCR, since its potential was beginning to become apparent; after all, legally it belonged to the company. Then again, the company might well have decided not to pursue PCR, simply because there were so many other pressing projects on everyone's agenda—and focus was the order of the day.

White set out explicit criteria for Mullis to meet and a timetable in which to meet them. Mullis had to choose someone to report to. After White declined to serve, Mullis picked Norman Arnheim, the new head of the Human Genetics department. Therefore, Mullis formally switched from Chemistry to Human Genetics.[45] White remarks:

> I was not convinced in June of '84 that the idea would work. I thought it might work, and if it did it would be really important. I wanted a written description of it and what he was actually hoping to do with regard to the experimental plan on a six-month basis. Actually, what he gave me was the invention disclosure that also turned out to be a workable experimental plan. I gave that to Erlich, who gave me some comments on it. Henry gave me the most explicit comments in July and August of '84 as to what were the criteria—the analytical criteria that PCR

should be judged on. It was one of the first documents commenting on the practicality of Mullis's idea.

The gist of it is that between June of '84 and August-September of '84, neither I nor Erlich nor Arnheim thought it was going to work by any data that we had seen up to that point. So in clear distinction from Kary's belief that the method had already been shown to work by June of '84, we did not feel that all the controls or all the correct experiments had been done and certainly not to the point where it could convince us or the reviewers of a peer-reviewed journal.

Mullis at that time was actively trying to get the thing to work. Mullis knew hardly any real biochemistry or molecular biology, but we were trying to help him learn how to do Southerns and sequencing or anything that would meet the criteria. Arnheim and Erlich were clearly interested in it and they were responsible for it in a sense, so they were trying to find out what Mullis was up to.[46]

Corey Levenson, who replaced Mullis as head of the oligonucleotide synthesis lab, observes:

At the beginning the response was sort of: why would you want to make all this DNA anyway? DNA probes were not popular because you have the sensitivity problem and you have to deal with radioactive labels that were expensive, and troublesome and hazardous. At the time we were thinking in pretty narrow terms about DNA probes. The DNA probe market was seen as that big. There weren't that many diseases that people were targeting. We were thinking of the utility of PCR in a very narrow area: looking for sequences of DNA that had some diagnostic significance. The money in biotech was in anticancer drugs. DNA probes were just like monoclonal antibodies, but that's not going to change the world.

Still, it is puzzling that so few people got excited about it. He was one of the only people to believe in it and to see some of the significance. In hindsight it's hard to explain that. If Cetus had just drawn the line and said, "We don't want to hear anymore about PCR; just quit talking about it or you're fired," he would *not* have quit thinking about it. He believed in it enough that he would have gone somewhere else and developed it if that's what it came to.[47]

Four

From Concept to Tool

By early 1984 Erlich, Saiki, and their colleagues had shown that their "oligomer restriction" method worked even on human DNA. They submitted and subsequently published a paper on the method and received a patent on it in 1987. Erlich and Saiki would be the first to admit, however, that OR was not an epoch-making invention. Erlich observes: "From my perspective we still hadn't achieved the objective of a simple test that would be appropriate for . . . medical testing because it still used radioactivity and . . . it wasn't that sensitive in the sense that we needed an amount of DNA to get out a signal in this test, [the amount of DNA needed] was more than was available in many clinical samples."[1] The process of developing the assay had nevertheless been extremely valuable because it had provided experience and skills in manipulating a variety of elements from reagent conditions to the complexities of enzymes; it had yielded a stock of experiences in strategic approaches to developing diagnostic systems; and it had established a productive collaboration.

On the PCR front, the responsibility for supervising Mullis fell on Norman Arnheim, who had come to Cetus in the fall of 1983 on a sabbatical leave of absence from the State University of New York at Stony Brook, where he was a professor of biochemistry. Arnheim was the first tenured professor to take a full sabbatical year at Cetus. He wanted to do some work that had practical implications in addition to his "esoteric" research on the evolution of complex "families" of genes.[2] Arnheim remarks that it would have been utterly unthinkable for someone like himself to be associated with industry during the 1960s, when he was in graduate school.

By the early 1980s, however, the anti-industry sentiment had declined significantly.[3]

Upon his arrival, Arnheim spent some time simply exploring and sampling several Cetus projects. He worked for a while with Frank McCormick on the *ras* gene, an oncogene, and a variety of antibodies to *ras* mutations; on Erlich's work on the genetics of HLA; and on some of the diagnostics technology. He met Mullis within the first week of his stay, when he visited his lab in order to be trained in gel electrophoresis techniques. Mullis enthusiastically explained his (non-PCR) strategy for developing an assay to distinguish between the normal beta-globin gene and the mutant allele. Arnheim remembers being skeptical about the approach but taken with Mullis's vitality and enthusiasm.[4]

Arnheim was duly impressed by the progress that Erlich and Saiki were making on the beta globin OR test. As it was easier and faster than Southern blotting, it appeared to be a significant technological advance. Since timing counted heavily in prenatal diagnostic tests, eliminating the use of Southern blotting could save several weeks. Erlich and Saiki's method, while promising, still had significant background problems caused by nonspecific annealing of the probe. Arnheim suggested using a "blocking" oligonucleotide. The idea was that once the annealing had taken place, but before the restriction enzyme was added, one could add an oligo that would "block" all of the probe molecules that were not specifically annealed to the beta-globin locus. The concept was a simple one: tie up the oligos that were just floating around with a much larger excess of the "blocking oligo." Once the probe was annealed to the appropriate allele and cut by a restriction enzyme, a very small, radioactively labeled oligo or di- or trinucleotide would be released and could then be run on a thin-layer chromatography plate. Experiments showed that the concept worked. Making it work well, however, required a large amount of total genomic DNA. This requirement brought the problem full circle—the scarcity problem remained unsolved.

Arnheim had little contact with Mullis during the year 1983–84. During the course of his stay there, Arnheim came to adopt the view of Mullis as someone with many "blue-sky" ideas and few data to back them up. At the beginning of the summer of 1984, Arnheim returned to Stony Brook to ponder a job offer to head

Cetus's Human Genetics department, which he ultimately accepted. At the end of the summer, the day he came back to Cetus, White outlined for him Mullis's PCR idea. This was the first time Arnheim had heard a cogent explanation of the principle of PCR. It would be prudent, White explained, to explore the approach further, even though, in his opinion, Mullis had not been able to get it to work. Specifically, White was impressed by Mullis's strategy of multiplying the target as a way to increase sensitivity. It might well be possible, White told Arnheim, to increase the target ten- or even fiftyfold, a qualitatively significant increase in sensitivity for diagnostic purposes.

During the summer of 1984, Arnheim had heard vague references to the Mullis method and remembers, "No one but Kary believed the data he showed . . . demonstrated what he said it did." White adds, "Any independent scientist looking at the data today would come to the same conclusion."[5] White thought that Arnheim and Erlich were the people at Cetus best able to help Mullis to either make the concept work experimentally or find any fundamental flaw that had been overlooked. Consequently, White formally asked Arnheim, in his new role as head of Human Genetics, to monitor what Mullis was doing, even though Mullis was officially still in the Chemistry department. Arnheim talked to Mullis (and Faloona) but was not convinced by the data and suggested they do some Southern blots in order to determine whether any of the multiple bands they were producing were what they claimed them to be. Arnheim remembers Mullis reacting with hostility to what he viewed as intrusive and unwarranted oversight of his work. In essence, he refused to cooperate.

For White, what was needed to determine whether PCR would work was independent analytical methods to establish whether the bands on the gels were actually the intended bands, to ascertain exactly *what* was being amplified. He didn't care if that required Southern blotting or sequencing or OR, or some other independent and scientifically demonstrable method. He wanted some reliable control. Because PCR's initial potential commercial value for Cetus lay in diagnostic tests, White felt strongly that the work had to be done on total genomic human DNA. Arnheim proposed going for the "gold standard"—a single-copy gene in a microgram of total human DNA. It was natural to choose the most developed experimental system available at Cetus, the beta-globin system. At White's suggestion, Erlich and Arnheim formed "the PCR group," which was staffed by scientists and technicians in Erlich's lab. The work of the group was codirected on a daily basis by Arnheim and Erlich. In addition, the group met regularly on Friday afternoons. Arnheim and Erlich (and others) fondly remember these Friday meetings—at least for the first year. Excellent interchanges over the data, took place, and the group members were open to constructive criticism, and goal-directed in seeking and proposing experimental strategies. For Arnheim, this was the first time in his scientific career that he had worked in a large group setting on an exciting, well-defined project; he found the brainstorming and feedback exhilarating. In his view, the experimental results generated from ideas that arose during the Friday afternoon meetings are testimony to the fact that PCR was made to work by means of a team effort. Mullis and Faloona had a standing invitation to these meetings, and usually one or both attended. Although Mullis was adamantly opposed to setting up the group, tension between Mullis and the others waxed and waned; during the first year (summer 1984 to late spring 1985), working relations were often collegial and productive.

MAKING IT WORK: A JOB FOR HIGH(LY SKILLED) TECHS

By the end of October 1984, White agreed to assign another technician to the project. The choice was Stephen Scharf, who had unquestioned technical ability with proven experience in performing Southern blotting. An added plus was that Scharf was a friend of Mullis's. There were three "technical" people centrally involved during the crucial development period of PCR—Fred Faloona, Randy Saiki, and Stephen Scharf. Their careers and personal trajectories present a contrastive range. None of the three had a Ph.D. Faloona, unambiguously at the bottom of the scientific hierarchy, maintained good relations throughout this period with the other scientists and technicians performing PCR-related experiments. His low status and affable personality enabled him to move from lab to lab with materials, (often) inconclusive results, and ideas without provoking any controversy. It was as if people didn't feel obliged to take a stand on Faloona; he was simply the bearer of reagents, autoradiographs, and news from Mullis as well as a conduit for returning the same. The praise and blame went to Mullis. Randall Saiki had moved through the ranks at Cetus and had been given a great deal of autonomy in defining his work. He was respected throughout the company. Stephen Scharf occupied a niche between Faloona and Saiki. Everyone involved expressed a high opinion of his technical abilities, and there was (and continues to be) a general confidence in his lab performance. Nonetheless, unlike the treatment accorded to Saiki, there had been a tacit reluctance to grant Scharf what amounts to full "scientific" credentials and the corresponding status, independence, and pay. Scharf is open in expressing his frustration over what he perceives to be a lack of recognition of his contributions at the time. For example, Scharf was not present at the infamous Monterey meeting.

> I was a Research Associate I at that time and the only people they would let go to those were Research Associate IIIs. That seemed to be stupid because ... it was the research associates who were the senior level of technicians at the time that they were actually doing the bulk of the work, and to not have them at a meeting where they were presenting the work seemed ludicrous but that's sort of what they did. I was very vocal about the fact that I should have been there.[6]

Of course, PCR was barely there, relegated to a part of a poster, overshadowed by big-ticket items and big egos.

By the summer of 1984 quite a lot was known about beta globin. Making the OR system work, however, continued to present problems: the signal-to-noise ratio remained poor. In the fall, Scharf was working full time on the beta globin–PCR project. In these experiments, cell lines of two types were used. They contained either the normal gene or the sickle-cell gene but not both. Scharf's experiments had specificity problems: PCR seemed to amplify both the normal allele and the mutation. There were several possible reasons for this confusing result. Perhaps the double signal had been caused by simple sloppiness in loading the gel. A precaution as simple as separating the samples by the width of the gel yielded better, although not perfect, results. Finding out why this and other inconsistencies remained demanded systematic (and time-consuming) exploration of experimental conditions. After a string of consistent results, Scharf remembers one experiment in which all the probes gave strong signals, including the empty control lane: "That was awful."[7] Though product was clearly being amplified and signals were being produced, they were simply too bountiful and were not always in the right places.

On 15 November 1984 Scharf was convinced that he really had "knock-out" experimental data. In his lab notebook he wrote, "IT WORKS!" The experiment clearly demonstrated that there had been amplification of the right-sized product. There was now no question that PCR was working and that specificity of amplification could be demonstrated. Seen from the lab bench—twenty months after Mullis's Eureka! experience—this experiment was the moment of triumph.

As soon as I saw that film I knew that it worked. As soon as I saw that signal I was very excited. . . . It was very clear what the significance of that was. . . . It reminds me of the guy who invented the laser in 1957 as a graduate student. The government came in and took all his notebooks and shut down his work; he had in his notebooks that the laser could be used not only for burning holes in steel but they might be able to contain a fusion reaction and this and that. He had seen all the applications at the time he had conceived of it. When I saw this result I said, this means we can do this and this and this, all these things. It would simplify cloning, it would cut down the signal-to-noise ratio, it would cut down exposure time to overnight from a week. Making a lot of DNA would make it easier to handle in the lab. This was a big deal. I remember when I precipitated these samples, spinning the stuff and looking and saying, my God there is a huge pellet of DNA in here. I told Henry, who said, "You can't see the DNA, it is a single-copy gene." Actually it had been amplifying not only beta globin but lots of other parts, making a lot of DNA. The probe was specific.[8]

Erlich had no doubts that something was being amplified (although he doubted the pellet was all DNA), but the question was, precisely what was it? How specific was this operation? These results, however encouraging, were still far from constituting convincing proof that the PCR group had worked out all the parameters required for a stable system. Nonetheless, demonstrating that the system could be made to work as it was supposed to, at least once, was a lift. Mullis, Faloona, and Scharf were now convinced that it was plausible and potentially powerful. Erlich and Arnheim didn't doubt that PCR was a powerful amplification technique, but they required more consistent data demonstrating specificity to be fully convinced of its viability. During the course of the following months, experimentation continued. Some experiments worked well and others didn't; some problems were solved and others remained. Although this lack of consistency was frustrating, there was nothing especially unusual about it—even in established systems, unexplainable problems crop up. However, as PCR-OR was not yet an established system, and as the goal was to produce a method reliable enough for commercial diagnostic tests, these remaining problems needed to be accounted for. The techs spent the winter sorting through the variables, seeking reliable standard procedures. For example, systematic work on varying the number of cycles of heating and cooling functioned as predicted (more cycles, more product). While there seemed to be some cross-reactivity between the amplified products, at least such cross-reactivity took place at a constant ratio. For example, using the same probe but decreasing the hybridization time in half-hour increments demonstrated that more time didn't increase the signal. Another example: a tenfold reduction in the amount of enzyme reduced the signal but made it more specific. For months, the group maintained a rhythm of a completed experiment every three or four days (a day to do the amplification, a day to do the OR and run the gel, a day to develop it).

Stephen Scharf reflects on the refinement of the PCR procedure:

> I think in a lot of respects when you work on a particular technique, it is sort of like a craftsman who develops an ability to do something—the more time you spend doing it, the finer control you get in doing something. I think a

classical example is DNA sequencing. The first time you run it, the gel looks terrible. It looks like not all the bands are there or the gel looks awful or the film looks awful or something, but as you do it more and more, you develop some sort of ability with your hands that allows you to get it to look better and better and at some point in time you make these beautiful gels and it's like a craftsman—you get this beautiful data. And I think that's sort of what was happening with PCR is that I was playing with it in many respects in terms of trying the various aspects of optimizing it that I could think of, and as a result, just doing it more and more, you get better at it. It starts to work more robustly. The reaction for the process is cleaner in the sense that the controls work the way they should work. You don't get surprises in terms of what's going on here or there. It becomes rigorous in your hands. It works the way you would scientifically think it would work when you plan an experiment.[9]

During the winter and spring of 1985, there was a consensus in the group that the results were encouraging, even exciting, but not yet quite definitive. At this juncture, Arnheim and Erlich decided to assign Randy Saiki—a consummate experimental practitioner—full time to the project. The investment paid off handsomely because Saiki eventually executed almost every experiment published in the first PCR publication. By this point the OR assay had been successfully worked out. A great deal of truly laborious and precise lab work was involved; nothing was automated and many of the frequent steps required constant monitoring over long periods of time. It was a tremendously labor-intensive undertaking, since it involved manually transferring tubes from a near-boiling water bath to one at 37°C, adding more polymerase, and repeating these steps over and over. Steve Scharf performed more of these repetitive tasks than anyone will ever have to do again, and has no regrets at seeing them taken over by a machine.

What had been achieved by spring 1985? The team members had demonstrated with reliable and quantifiable data that they could amplify genomic DNA hundreds of thousands of times, thus amplifying the target they were aiming at. The precision of the result was ascertained by quantitating the amount of a specific sequence radioactively and using the oligomer restriction assay. The data showed that not only had there been exponential amplification but that it was the beta-globin gene that had been amplified. Given the objectives of the experimental system, and given the fact that OR could be used to detect a specific product, the fact that other DNA was also being amplified did not pose an insurmountable problem. The system's specificity was good enough for diagnostic purposes.

GOING PUBLIC

During the spring of 1985, Arnheim, White, and others went on a series of "road trips" to large companies such as Kodak and Smith-Kline to present Cetus's diagnostic program. Cetus had decided to sell the diagnostic technology it was developing in monoclonal antibodies and DNA probes. Although monoclonals were the hot item at that point, amplification technologies were usually mentioned as part of the presentation. The purpose of these presentations was to attract investments and possible partnership deals from these larger corporations by demonstrating that Cetus had a broad base of technologies and that the scientists at Cetus were making discoveries that *could* have valuable applications in the commercial world. The presentations contained few details, since they were meant more to entice investors for future developments in "cutting-edge biotechnology" than to sell any particular product. Arnheim remembers talking in vague terms during the spring about work under way on technologies for "increasing the target." In one presentation, Arnheim remembers an acquaintance in the audience asking questions that indicated he was getting close to grasping the concept of PCR. White, Price, Arnheim, and Erlich began to realize that they ought to start thinking seriously about publishing their results or else risk losing credit—scientific and commercial—for PCR. One risk was that even if someone merely added a short section on PCR to the end of a paper on another topic, he or she could well receive the credit for priority. Risks such as this needed to be taken seriously. The first PCR patent application was filed on 28 March 1985.[10]

A significant payoff of the road trips was a joint venture agreement between Cetus and the Perkin-Elmer Corporation in December 1985 to develop instrument systems and reagents for the biomedical research market. Although PCR was not one of the initial goals of the agreement, as PCR began "exploding" two years later, the reagent and thermal cyclers produced by the joint venture soon became the primary focus of their activities as well as a commercial success.[11]

In March or April 1985—two years after Mullis first "saw" PCR—Arnheim received the abstract forms for the annual meeting of the American Society for Human Genetics (ASHG) scheduled for late October. The deadline to submit abstracts was 30 June. After some informal discussion among the R&D leadership, a consensus formed that the upcoming meetings presented a good opportunity to present PCR in its application to human genetic disease. In May a meeting took place in Jeff Price's office with White, Erlich, Arnheim, and Mullis to formulate a timetable for publishing. Everyone but Mullis concurred that an abstract should be submitted and a paper published after the meetings. They agreed that the (fifteen-minute) talk at the meeting would not disclose precise experimental conditions but only results, employing generalities to portray their significance. White observes:

> By the spring it would have been clear that the thing worked to Henry's whole department plus to a wider group just by word of mouth. There would be another twenty-five who probably had heard something about it and just wouldn't believe it. . . . Erlich and I and the patent attorney argued that we should publish the method and the business person argued that we should keep it a secret. We were in discussions with Kodak about a long-term diagnostics relationship and if we published it, it would put the method in the public domain and therefore be of less interest to Kodak—it's something they would have exclusive rights to. We would argue that we would have exclusive rights to it anyway, and who knew if the Kodak deal would go through or not—the method was too important to keep a secret. So from June on there were some heated arguments over the writing up of that work.

Mullis didn't want to publish it. I viewed this as procrastination.[12]

From this point forward, charges of procrastination and countercharges of undue speed—equated with misappropriation—became more and more frequent. The plan was to have two papers, one theoretical and one applied. The group unanimously agreed that Mullis should write up a full description of the fundamental PCR concept using the experimental data he had on the pBR322 plasmid. The "applications" paper, on the beta globin–PCR system, would then be submitted after Mullis's "fundamentals" paper. It seemed to Arnheim and White that Mullis already had, or was close to having, sufficient data to convince a reviewer that the concept was valid, i.e., that he was amplifying a defined and identifiable segment of the plasmid. Consequently, it seemed entirely feasible for Mullis to submit a paper outlining the fundamentals of the method before the October meeting (and possibly have it already accepted for publication). His article would appear first. Then it could be followed quickly by others.

Mullis was categorically opposed to these plans. At first he argued for keeping PCR as a trade secret with no publication at all. White recalls that Mullis proposed selling tubes containing reagents—the mixture of reagents would remain a secret—into which one had put DNA; people would be amazed, Mullis argued, at the end when they saw the huge amount of DNA produced.[13] White pointed out that this was not a plausible plan because, among other reasons, it would be perfectly possible, and not terribly difficult, to reverse-engineer the kit. Further, other people at Cetus were beginning to work on various applications of PCR, and such work would inevitably lead to other papers that would require the publication of the experimental conditions in the Methods section. Besides, Cetus had already filed a patent in March 1985, and the eighteen-month clock before mandatory public disclosure was already running. Eventually, everyone—including Mullis—agreed that they would submit the abstract. Instead of Erlich and Arnheim's jointly taking responsibility for writing the applications manuscript, it fell on Arnheim because Erlich was occupied with other work.

In July Mullis and Arnheim had a long and unproductive conversation over Mullis's work plan for the coming year. Arnheim recalls:

> At the end of the year Kary had to present the work for the year to a committee of five people: myself and two that he picked and two that I would pick. That was it. Compared to performance reviews for others it was the least restrictive of any scientist in the company but far too restrictive in his opinion. . . . Kary wanted to do what he wanted to do, when he wanted to do it, and didn't want anyone to tell him anything about anything. The big negotiations were that he would attend the PCR meetings and that at the PCR meetings he would report on his work. He didn't want to have to report on his PCR work. Period. He felt that his contributions were so valuable that he should be left alone to do his work as he wanted to do it. Period. . . . He believed . . . Henry and I were policemen. We "stole" his work. We were suggesting experiments that were stupid. OR was a failed method. White never adequately supported him; he was a bad friend. I "bailed," as my daughters say, and left it in Tom's hands.[14]

Arnheim had finally decided to accept a position as head of the biology department at the University of Southern California (USC). He left Cetus in August 1985 not because of Mullis but because he was spending more time doing "practical" things than he had intended; had his "esoteric" work fit into the frame of Cetus's commercial goals, as Erlich's did, he might well have stayed on. At USC Arnheim went on to pioneer the analysis of DNA in a single cell using PCR. He continued to consult for Cetus on a regular basis. Mutual respect was declining, the tide of trust was ebbing, the clock was ticking, the page was turning—multiple metaphors apply.

Crescendo: Fall 1985

What was the state of PCR at the end of the summer? White remembers:

> It was spreading like wildfire and everybody believed the thing worked by that time and people were using it for

all kinds of stuff. I think in September '85 Mullis actually gave a company-wide seminar. I remember distinctly that in that seminar—by that time, of course, he had convincing data—he also had a bunch of other ideas that were whacko new ideas in a certain way and his seminar style was always so eccentric that people got up and walked out on his seminar. Mullis always bitterly remarked to me afterwards that, even when he felt he had proven it, other people were still walking out of his seminars. I knew Kary had friends outside the company—friends that we had in common—artists, for example. I was surprised to hear from an artist friend of ours that Kary described PCR to him one night over dinner. I knew we had mutual friends at Genentech and Chiron and all over the place and sooner or later he might describe it to them. Some of the consultants were known to be somewhat predatory about publishing ideas that came out of the scientific retreats as their own ideas—this had happened on a number of occasions with no credit to where the idea had originated—I felt we were in a dangerous situation.[15]

PCR had to go public. Although attentive to PCR's progress, White too was devoting the bulk of his time to the more high-profile projects at Cetus. It is fair to say that PCR was emerging as an effective tool, a facilitating technology, and that neither Cetus scientists nor management thought of themselves primarily as toolmakers.

During the summer Saiki performed a series of "aesthetic" experiments to prepare the data for publication in an elegant form. He was chosen to make the presentation of PCR at the ASHG meeting, his first public talk at a major scientific meeting. That summer *Scientific American* published an article by Mandelbrot on fractal geometry. According to White, "Mullis would bring in beautiful prints of Mandelbrot patterns every morning—using up a whopping amount of computer time instead of writing. His personal life was still very difficult."[16] White gave Mullis an absolute deadline—November 1—by which to complete the "fundamentals" paper, after which Cetus would submit the "applications" paper for publication. Once Saiki presented PCR at the meeting in

October, Cetus would be under the commercial and patent gun to get PCR into the literature.

PUBLICATIONS: ANALYSIS AND SYNTHESIS

The "applications" paper was submitted to *Science* on 20 September 1985, was accepted on 15 November 1985, and appeared on 20 December 1985. *Science* has several rubrics; the five-page piece was published as a Research Article, the most prestigious category. Both the title—"Enzymatic Amplification of Beta-Globin Genomic Sequences and Restriction Site Analysis for Diagnosis of Sickle Cell Anemia"—and the listing of authors—Randall K. Saiki, Stephen Scharf, Fred Faloona, Kary B. Mullis, Glenn T. Horn, Henry A. Erlich, Norman Arnheim—are significant. The use of a conjunction in the article's title underlines the overlap still existing at the time between Mullis's concept, the OR experimental system, and the (commercial and research) context at Cetus.

The abstract announces the successful application of two methods for a rapid and highly sensitive diagnostic test for sickle-cell anemia. The first, PCR, "involves the primer-mediated enzymatic amplification of specific beta-globin target sequences in genomic DNA, resulting in the exponential increase (220,000 times) of target DNA copies."[17] The second, OR, rapidly detects the presence or absence of two beta-globin alleles. Two techniques, one experimental system, one context (genetic and infectious disease detection). PCR—unambiguously attributed to Mullis and Faloona—is described as a powerful amplification technique: "repeated cycles of denaturation, primer annealing, and extension result in the exponential accumulation of the 110-base pair region defined by the primers."[18] The article's first figure legend presents the basic elements of PCR not as a general method but (as agreed) as a component of a method for amplifying human genomic DNA, with the purpose of using another technique (OR) to detect the presence or absence of the beta-globin alleles.

The section entitled "Diagnostic Applications of the PCR-OR System" lauds the greatly improved rapidity (from "several days" to ten hours) the method provides for a prenatal diagnosis of sickle-cell anemia, its simplicity (compared with the Southern transfer procedure), and its sensitivity (it works even with degraded DNA). Further, the method is not dependent on the use of radioactive probes. Finally, the article clearly indicates the limitations of the PCR-OR method—the requirement of a restriction site in the vicinity of the locus of interest, the need for genetic linkage studies. PCR, the article states, is a method of analysis.

The article concludes: "The ability of the PCR procedure to amplify a target DNA segment in genomic DNA raises the possibility that its use may extend beyond that of prenatal diagnosis to other areas of molecular biology."[19] This laconic understatement self-consciously echoes the famous ending to Watson and Crick's 1953 article in *Nature* announcing the double-helix structure of DNA: "It has not escaped our notice that the specific pairing we have postulated immediately suggests a possible copying mechanism for the genetic material."[20] Both predictions proved correct.

Synthesis

In December Mullis submitted his paper to *Nature*. He neglected to include a cover letter making clear exactly how his paper differed from the *Science* paper. The paper was rejected as merely technical and unoriginal. White—and everyone else at Cetus involved with PCR—was stunned. White hurriedly wrote a cover letter with Mullis; Mullis resubmitted the paper to *Science*. It was rejected.

Mullis remains extremely bitter about the publication history of PCR. He feels strongly that credit for his invention was stolen from him. He assigns primary responsibility for this purported villainy to Arnheim and Erlich, but he considers Saiki and White tainted by their complicity.[21]

> I had sent it to *Nature*. I had said there is a *Science* paper that's already in press. *Nature* asked to see the *Science* paper. By then the *Science* paper had gone through many different editions, and every time Norman had added more and more detail about PCR, and more and more that sort of sounded like they had really developed it themselves, that it wasn't somebody else's technique that they used. The paper that Normal was writing was really supposed to be about oligomer restriction and it was sup-

posed to have used amplification, but it was supposed to
be two separate things. By the time he got it in there, two-
thirds of it was amplification.

[*Nature*] didn't call me because I wasn't one of the good
old boys. . . . *Science* rejected it too. . . . [They wrote back]
It's a technique, is all. . . . That just sort of totally knocked
me for a loop. I said, "My god, here Erlich and Arnheim
and Saiki are going to get away with their little trick."
They've got their paper there.[22]

Dismayed and scrambling to rectify the situation, senior R&D
people at Cetus looked for a place to have Mullis's paper published
rapidly. Shing Chang suggested submitting Mullis's twice-rejected
paper to a volume of *Methods in Enzymology*, a respectable "methods" journal, albeit one not carrying the prestige of *Science* or *Nature*. The editor, Ray Wu, a close friend, was likely to be accom-
modating. White arranged with Wu for rapid consideration of the
manuscript. It was accepted in May 1985 with an anticipated early
1986 publication date. However, a series of absolutely atypical delays put off publication of the issue containing Mullis and Faloona's article for almost a year. What *should* have been the first comprehensive presentation of PCR—"Specific Synthesis of DNA *in
Vitro* via a Polymerase-Catalyzed Chain Reaction," by Kary B.
Mullis and Fred A. Faloona—belatedly appeared in a 1987 volume of *Methods in Enzymology*. By the time it appeared, its message
was well known: a rapidly growing base of experience using PCR
(transmitted informally both within Cetus and beyond) was circu-
lating about the concept, its pertinent technical specifications, and
its potential applications and variations.

At about the same time that Wu was considering the manuscript, Frank McCormick of Cetus had contacted James Watson
at Cold Spring Harbor and strongly urged him to have Mullis
present PCR at an upcoming May 1986 symposium on "The Molecular Biology of Homo Sapiens." Watson agreed. Mullis made
the most of the opportunity when he presented his talk at Cold
Spring Harbor, the first "invited lecture" he lists on his vita. "I . . .
have a refreshing kind of style, like it's not drab at all. It was easy
for me to get their attention and to show them that it was definitely
a significant thing. . . . There was an overwhelming response.

There were a lot of people who wanted to know how to do it. There were all kinds of people who were very excited about it."[23] As a result of his performance, Mullis became publicly associated with PCR in a major arena of scientific visibility. The proceedings of the symposium—the first PCR publication with Mullis as first author—were published as a volume of the series Quantitative Symposia in Molecular Biology in late 1986. The Cold Spring Harbor publication, though prestigious, was not considered to be a "primary" publication, and received substantially less attention than it warranted. In subsequent years, many scientists working on PCR appear never to have read the Cold Spring Harbor article, entitled "Specific Enzymatic Amplification of DNA in Vitro: The Polymerase Chain Reaction," by K. Mullis, F. Faloona, S. Scharf, R. Saiki, G. Horn, and H. Erlich. In it the authors clearly present the concept of PCR, the basic variations in deploying it, and the initial applications that demonstrate its power, versatility, and promise.[24]

The form and contents of the Cold Spring Harbor paper were the result of a great deal of discussion and negotiation about exactly what should appear where and in what detail. There were priority considerations (some of Mullis's material was not included so that it could be reported in his own paper), patent considerations (how much detail should be included on variations such as the use of primers that had promoter sequences attached), and debate about whether to report promising, but not yet definitive, results (work on the thermostable *Taq* enzyme). By common accord, Mullis was made the first author. Mullis wrote the first part of the article, adopting a style that boldly lays out the general significance of PCR as an incredibly powerful synthetic tool. Saiki and Erlich wrote the second part of the article, in which they sketch the horizon of possibilities for the analysis of genomic sequence variation as well as its initial applications.

Mullis presented PCR as the culmination of a line of technological advances in DNA synthesis. He traced its lineage to the 1970 discovery of restriction endonucleases, which made possible the isolation of discrete molecular fragments of naturally occurring DNA, the precondition for molecular cloning, achieved in 1973. *De novo* synthesis, even with automatic equipment, Mullis underlined, remained technically difficult, and, most important, "is not capable of producing, intentionally, a sequence that is not yet fully known." As long as it was only possible to synthesize what one already knew, then analysis always preceded synthesis. PCR provided the means to reverse that order. PCR offered

> an alternative method for the synthesis of specific DNA sequences.... The PCR method is completely general in that no particular restriction enzyme recognition sequences need be built into the product for purposes of accomplishing the synthesis. Furthermore, the PCR method offers the convenience of enabling the final product, or any of the several intermediaries, to be amplified during the synthesis or afterwards, to produce whatever amounts of these molecules are required.[25]

These claims unambiguously distinguish PCR from any particular use to which it had been or might be put. DNA synthesis is no longer dependent on nature's biological constraints. PCR turns scarcity into abundance.

ACCURACY: FINDING A BETTER POLYMERASE

The last major piece required to make PCR into a machine was the introduction of a more efficient, specific polymerase. The following account of the identification, purification, and introduction of *Thermus aquaticus YT1* DNA polymerase, or *Taq,* into the PCR system (again) juxtaposes a version culled from Cetus's senior scientists with a counterversion from Kary Mullis.[26] Mullis was the first to propose (during the spring of 1985) using a thermostable polymerase, i.e., one capable of withstanding the temperatures required by the denaturing step without being deactivated. The impetus to look for a different enzyme emerged from the material requirements of the first instrument designed to automate PCR. "Mr. Cycle," as it was called, required heating a tube containing a sample to 95°C to denature the DNA, and then cooling the tube to 37°C, to optimize primer annealing and to begin polymerization.[27] While crude in its initial form, this automation was nonetheless enormously laborsaving—no one had to stand over the tubes adding polymerase at each cycle.

The Klenow Fragment

The so-called Klenow fragment (roughly two-thirds of the whole polymerase) of *E. coli* POL I (DNA polymerase) originally was used in PCR because it was well studied and commercially available. Arthur Kornberg (and eventually his son) had done a good deal of scientific work characterizing its properties. Essential components of what was considered state-of-the-art knowledge about the *E. coli* POL I polymerase are now known to have been misunderstandings. *E. coli* actually has at least three polymerases. POL I was not responsible for genomic replication but was actually a repair polymerase.[28] The history of the scientific understanding of polymerases had nothing to do with biotechnology. *E. coli* was a model organism of choice in contemporary genetics, and for that reason the enzyme was commercially available. It had been developed for an entirely different market, one whose experimental systems, techniques, and goals required much smaller quantities of POL I than PCR required. The polymerase's properties happened to do—or could be made to do—more or less what Cetus scientists required it to do. Had the properties of other polymerases been explored, the scientific and technical strategies employed would certainly have been different. Klenow became the standard against which efficiency was measured.

Gelfand, who had been asked by White and Price to attend the PCR meetings as an advisor with expertise in enzymology, completely agreed with Mullis that this was a potentially promising area to explore. In Gelfand's eyes, a literature review revealed not only how little was known about such polymerases in general but, more important, how little of what was known about their properties was relevant to the questions Cetus wanted answered. The two specific functional characteristics needed in an enzyme were polymerase activity around 65°C following exposure to a temperature of 95°C. The former quality was evident (the organism grew at that temperature; hence, the enzyme had to be active at that temperature), the latter unknown. No one had previously looked at these questions because there had been no reason to undertake that kind of study.

There was a group working on thermophilic enzymes in Cincinnati, and another group in the Soviet Union.[29] The Soviets had a more sustained interest in the area, producing a paper on a different species every year and a half. Although there was general agreement within the PCR group that it was worth purifying a thermostable polymerase, a decision had to be made about which one to start with. There were a number of theoretical or biological considerations to be taken into account on which the available literature cast no light. Whereas it was reasonable to suppose that these organisms possessed an enzyme that would function at 65° to 75°, it wasn't necessarily a given that they had a pure protein that would function as a pure protein away from other cellular components at this temperature range. It was beginning to be understood that polymerases operated within a complex cellular milieu and that the enzyme might well require multiple elements in order to function. It was entirely possible that if some of those factors were purified away, the polymerase would simply not function properly. The Cetus scientists knew what they wanted the enzyme to do—survive the denaturing step and perform the replication function at a lower temperature—a capacity it had obviously not evolved. Whether or not this polymerase, or any polymerase in this group, could be made to meet these requirements was simply unknown.

The task was straightforward: purify the polymerase and explore its properties in an artificial milieu. Although there was a large protein chemistry department at Cetus, no help in purifying *Taq* was forthcoming from those quarters because company management, specifically Robert Fildes, was adamantly opposed to "wasting resources on this non-therapeutic project," an opinion he steadfastly reaffirmed as late as 1990.[30] Both Mullis and White approached the protein chemists about working on the project, but the chemists were under heavy pressure to advance the therapeutic projects and simply could not spare the time. Mullis remembers:

> Then I tried to get somebody at Cetus to make a thermophilic polymerase for me, and asked almost every protein chemist at Cetus and nobody had time. They were all busy doing things, like one of them was busy trying to improve

the sensitivity of horseradish peroxidase detection of DNA. I said, "This method here can make 100,000, a million copies of DNA. We don't need sensitivity of detection." His lab had spent about a year trying to improve it by a factor of two, and I said it wasn't necessary any longer. He didn't want to make a thermophilic protein.[31]

Mullis refused to do the work himself; White knew that Mullis had the required biochemical knowledge and abilities to do the work. The issue was raised at a PCR group meeting. Gelfand, while not a specialist in DNA polymerases, had purified a variety of proteins (restriction endonucleases, ligase enzymes) during the course of his career and was generally familiar with protein chemistry. He responded that his lab had the necessary equipment, and he encouraged Mullis to get to work, offering to assist him as required. He remembers saying something to Mullis like "just come to the lab, use whatever you want to use—it's yours. And if you need some assistance, I'd be glad to show you how to run a column or pour a column and how to assay fractions."[32] That help was never asked for.

The PCR group was getting more and more frustrated waiting for Mullis to purify the enzyme. The group decided to give Mullis one last chance and then to have Gelfand proceed with the work himself. After another few weeks of waiting, Gelfand had Cetus request the available strains of thermophilic enzymes from the American Type Culture Collection. Next, the protein needed to be purified. The purification work was done by Gelfand and Susanne Stoffel, a Swiss national with a background in purifying enzymes who had been hired at Cetus in 1978. The purification and characterization took approximately three weeks of overtime work. Gelfand reports: "[I]t worked like a charm! There was a single band on the gel from human genomic DNA for beta globin. First shot out! The holy grail had been achieved. It had worked better than even we had fantasized. It was just so striking."[33] Gelfand is positive that Mullis was given at least a third of the purified enzyme within a day of its being available. Mullis's version casts events differently: "By June [of 1986], the [*Taq*] polymerase [had been purified]. They brought all of it to Randy Saiki instead of to me, and that pissed me off. . . . I sent Fred down. I said, 'Just go down and get half of it.' We set up a couple of experiments that day, and by that afternoon we knew that it would work, which I had thought it would."[34]

Taq worked with the existing instrumentation: hence, even if it had been as "sloppy" as Klenow, it would still have had large advantages. But *Taq* was not as "sloppy" as Klenow; far from it. It was a great deal more specific and produced a dramatically higher yield of intended target; the 99 percent of background product amplification from Klenow was gone. Although the greater specificity was not a total surprise, the degree was striking. As Gelfand puts it, "We didn't realize how 'bad' Klenow had been because Klenow was so much better than anything before it."[35] Whereas Klenow had yielded a 200,000-fold amplification of the target, *Taq* achieved a quantum leap over that quantum leap.

MULLIS LEAVES CETUS: SEPTEMBER 1986

Mullis absolved himself of any responsibility for the publishing debacle. As recompense, he demanded to be the first author on any other PCR paper that came out of Cetus for the next five years. This demand as well as the continuing priority fights (and a patent dispute) added yet more irritant to highly inflamed wounds. Mullis recounts:

> I was really mad at Tom. . . . It was the shabby goddamn way I was treated there. . . . If I am saying something about [Norman] that's slanderous or libelous, that's between me and Norman. That's not between me and this company. . . . But Tom said, "Don't talk about him anymore in public." I said, "I'll talk about him when I damn please." He said, "Well, if you value your job here, you won't." We talked about it a little more, and it was clear that I wasn't happy in that situation. . . . Tom said, "If you want to leave here, I'll give you five months' salary and stuff." I thought, "That's exactly what I want—Thanks." And I left.[36]

The rights to PCR belonged to Cetus Corporation. As there were no products based on PCR at this point, the value to Cetus of PCR was as a (potential) contribution to the Kodak diagnostics program. In the spring of 1986, Mullis received a bonus of $10,000. Scientists at companies typically receive only the nominal sum of $1 for an issued patent. Hence, one can compare, as White does, Mullis's $10,000 to the $1 that Erlich, Saiki, and others received for their contributions, or to the $300 million paid by Hoffmann-La Roche to Cetus for the rights to PCR. The bonus Mullis received was the largest ever paid to a Cetus scientist for an invention or other contribution to a product. It was also the first such bonus that had been paid since Fildes's arrival at Cetus, since he had ended the previous cash bonus program for scientists (bonuses continued for management). In early 1986 there was only one publication on PCR, the Kodak and Perkin-Elmer programs were just beginning, and there were no products. The fundamental patents did not issue until June 1987, and the first PCR reagents (*Taq*)and thermocyclers were commercialized in 1988, almost two years after Mullis quit. Thus, one could hold the view that all of the crucial work done to develop the value of PCR as a research tool and diagnostic method was done by others at Cetus after Mullis left, and that after 1986 he contributed nothing to this value. Finally, had he stayed on at Cetus, he would have realized more of the value of PCR, via stock options and salary increases, as did those who commercialized the first products.[37]

Five

Reality Check

In the early 1980s there were no molecular DNA tests for genetic diseases or genetic identification. Informed opinion in the industry held that the commercial diagnostic world was likely to be highly unreceptive to genetic tests. By the mid-1980s there was only one monoclonal antibody available to detect one particular allele predisposing people to a disease. Even in the case of diabetes, whose incidence in the U.S. population was high, the disease progression was not well understood; there would therefore be few therapeutic advantages to presymptomatic diagnosis. Forensics and paternity testing were other possible areas of commercial interest for genetic testing. The total value of forensic services was small. Paternity testing was a bigger market, but existing serological methods were less expensive than DNA tests. In the short run, the common wisdom held, human genetics offered few prospects for commercial success.

PCR opened the door for an extraordinary proliferation of knowledge in many areas, none more expansive than the HLA system. Although there was evidence from the serologic data that HLA genes were highly polymorphic—and that this variation had functional importance—mapping and sequencing was a tedious affair.[1] Furthermore, it was difficult to evaluate variation because insufficient data were available to establish a population data base. Although the scientific interest of having such a population-based framework was becoming increasingly evident, building one would be a lengthy process, given the available technology. PCR offered a means to increase greatly the speed and accuracy of all such work.

Henry Erlich's lab began working on a number of changes in the beta-globin system to adapt it to the HLA DQ alpha gene. The first concern was the size of the target. The beta-globin fragment was 110 base pairs long, at the time about the longest fragment that had been successfully amplified. Some innovative changes in the reagents, introduced by Stephen Scharf, improved PCR's ability to amplify larger fragments, improving the range to between 300 and 400 base pairs. Fortuitously, the HLA Class II genes were almost perfectly suited for PCR because all of the polymorphism was localized to a small fragment (the second exon in the gene, which stretched over some 300 base pairs). Even in the early days of PCR before *Taq*, it was possible to amplify the polymorphic region of the HLA DQ alpha genes and then to clone and sequence the alleles. Although genomic libraries would technically no longer be required thanks to PCR, the lab often continued to clone. While the cloning step required more work, it also yielded a more accurate picture of variation because in a heterozygous individual both alleles could be separated and sequenced. Using these procedures, genes from a large number of people could be easily isolated, cloned, amplified, and sequenced, producing an exponential growth of data. The region's polymorphism was rapidly charted; over the next several years, scientists around the world identified 90 percent of the currently known variations.

Once the extent of allelic variability became clear, it was possible and logical to design simple, rapid typing procedures. For population studies, it was not necessary to sequence each individual. The lab developed a typing system, employing oligonucleotide probes, based on the known pattern of sequence variability. Since practically all of the new alleles were simply different combinations of the older polymorphic segments, it was easy to design a panel of probes with which to type new variants. A panel of a hundred samples can be genetically typed through a pattern of probe binding (now computerized), allowing rapid identification of new alleles. These typing advances led not only to a vastly more masterful understanding of the variability of this key component of the immune system but eventually in 1990, to an HLA-based DNA forensic test.[2] This work opened the door to the mapping of human genetic variability. It is no exaggeration to claim that PCR is a fundamental tool that makes feasible such megaprojects as the Human Genome Initiative.[3]

John Sninsky joined Cetus in May 1984, just at the time the now famous articles by Robert Gallo and Luc Montagnier announcing the identification of retrovirus involvement in AIDS were published in *Science*. Sninsky left a position at the Albert Einstein College of Medicine in New York, where he was an assistant professor of microbiology and immunology with a joint appointment in molecular biology. He was working on the hepatitis viruses, specifically the human hepatitis B virus, as well as other related animal viruses. Sninsky was no stranger to Cetus, having consulted for the company during a four-year postdoc period (1976–80) at Stanford. Sninsky recalls:

> New York was a hard place to live—exciting but a hard place to live. I couldn't buy a house very readily, my car was broken into on frequent occasions. It is not a great place because of the social/economic status of much of that city—kind of depressing. I really liked the people at Einstein, but one of the things that concerned me is that even though I did a lot of teaching and got grants to cover all of my salary and all of my research, my overhead was very, very high, and I was asked to sit on committees and teach, but not really be reimbursed or compensated in any way, and all efforts to get 10 or 20 percent of my salary covered basically fell on deaf ears.

Upon his arrival, Sninsky proposed working on HIV. There were several reasons to support his request. A Cetus subsidiary, Cetus Palo Alto, was already doing work on cytomegalovirus (CMV), and there were other viruses not well served by existing tests. It was already known that the "AIDS virus" was hard to cultivate in cell culture, making it a likely candidate for a nucleic acid test. However, marketing argued that AIDS was a disease restricted to homosexuals and consequently wasn't going to be a significant financial opportunity. Sninsky brought the proposal to Jeff Price and Tom White, who thought that scientifically it was a reasonable avenue to pursue. In early 1985 they transferred a talented technician, Shirley Kwok, to work with Sninsky.

Shirley Kwok's self-description in many ways follows the stereotype of the "model minority." The daughter of first-generation Asian-American working-class parents, she grew up in San Francisco, attending the elite Lowell High School and the University of California at Berkeley, where she majored in bacteriology. She also represents the entry of a new type of technician, capable of inventing new skills—such as PCR.

INTERVIEW: SHIRLEY KWOK

SHIRLEY KWOK   After four years at Berkeley, I decided that I wasn't all that interested in school—the whole atmosphere at Berkeley wasn't all that appealing. It was really competitive. You were just another face in the crowd. And so I didn't think of graduate school at first. I wanted to go out and get a job for a while and think about what I really wanted to do with my life.

Two months after I graduated, I got a position at Cetus. The placement center was advertising an entry-level position as part of an assay group, screening different isolates for improved production of antibiotics. I was glad to have a job because the unemployment rate was really high. I worked in that lab for two years and then moved on to work for a series of scientists including Jeff Price. As Jeff's administrative responsibilities grew, he became less involved with bench science. One day, he mentioned that Tom was looking for help and asked if I would like to join him. When I said yes I was transferred over to Tom, who was working on B_{12} improvement. Afterwards we started getting involved with recombinant DNA and molecular biology. Tom was trying to learn it at the same time. I certainly had no experience. But that was great because it gave me an opportunity to try out new techniques.

One of the good things about Cetus was that there were opportunities in different areas and every couple of years I would move on. It wasn't stagnant. Tom and Jeff helped my confidence grow. And subsequently with David Gelfand and John Sninsky as well, I feel I've been really fortunate in my career.

PAUL RABINOW   You were working directly with Tom?

SK   I was in a special position with Tom. He was also moving up in the corporate ladder and gaining more responsibility. I think he negotiated with Ron Cape at the time to carry out his own special project—to look at the molecular evolution of fungi. I was the person he had asked to do this special project. I was not really in the mainstream of other projects at Cetus, and I felt isolated from the rest, although it was a nice isolation. I really didn't care to be part of the larger groups from what I heard about some of the politics going on. It gave me the opportunity to learn the recombinant DNA technology, learn at my own pace, and yet at the same time generate useful data. That lasted for about a year and a half. After that I had become fairly proficient with the technologies. Eventually, Tom had a hard time justifying keeping me to do his special project and thought that for my own career advancement I would be better off if I helped on another project. That's when I started helping David. I spent a year and half or so with him, doing a lot of cloning. In 1984 John Sninsky came aboard and I began working with him on diagnostics. I worked for at least a year cloning a syphilis gene. That was about the time when Kary discovered PCR. At that point, because of the new amplification technology, John thought, "What better target than HIV?"

PR   What was your role in developing the project?

SK   By the time we initiated the project, HIV was identified as the agent associated with the acquired immune deficiency syndrome. John had asked me if I would be willing to serve as the project leader. I felt honored but at the same time nervous. From a scientific point of view and for my own career development, this virus had much to offer. However, the potential risk of getting infected had to be considered, particularly in light of my two young boys. After lengthy discussions with my husband and thinking about the precautions that would be in place, I decided the health risk would be extremely low. I was also nervous about heading the project. I'd been a technician up until that point and although I had worked quite independently, I wasn't quite sure what was really expected of me. After getting assurance from John we would be working together as a team, I decided to take on the new challenge.

At the point where I got the project I was a research associate. By the time I finished developing the HIV assay, I was already promoted to associate scientist, a Ph.D.-level position. I was promoted to director of the Infectious Diseases department last year (1994).

P R   The story you're telling me is so idyllic. Wasn't there any racism and sexism?

S K   I wouldn't say so. No, we're dealing with a small subset of people in our daily lives—at least in this facility. Maybe I'm just sheltered and don't see what's really out there.

Sninsky and Kwok tried obtaining characterized viruses from their colleagues in governmental institutions in both the U.S. and France to no avail; all of Cetus's requests to the Pasteur Institute and the NIH were rejected. This stonewalling obliged Sninsky to embark on building his own control template for model experiments. Once that task was under way, Sninsky was able to establish a collaboration with Bernie Poiesz at SUNY, a collaborator of Gallo's who not only had experience in HTLV-1 but had begun looking at AIDS patients. Poiesz provided Cetus with its first clinical samples.

The challenge that Sninsky and Kwok faced was the detection of a small number of copies of what seemed to be a highly variable virus. Randy Saiki had already shown that a small number of copies could be detected in complex DNA. Sninsky was soon able to detect one infected cell in 250 uninfected cells. This technical feat, while impressive, still lacked the power required for a reliable HIV test. As it was already known that the virus mutated within and among patients, a major challenge was to find a common region to target. The first step was obtaining sequence data from multiple viruses. It seemed reasonable to concentrate on the functionally important regions of the virus (as they were becoming known). Then the lab could begin designing probes for those regions. Little was known about the ability of the polymerase to accommodate mismatches in a PCR reaction; a small number of mismatches could preclude amplification. In addition to sensitivity issues, lentiviruses were known to display a marked heterogeneity in viral DNA, most notably in the *env* [envelope coding] gene. In order to design a successful DNA-based test, it would be crucial to address this variability, probably through the use of multiple primers. In all probability, a useful test would not only have to be highly sensitive, it would also have to be capable of locating multiple targets.

In July 1985 Sninsky and Kwok circulated a proposal for the detection of AIDS-associated virus(es). The principal research questions were (1) Do the AIDS-associated viruses reside in the immune-system cells as DNA copies prior to the onset of symptomatic AIDS? (2) Can PCR be used to amplify viral DNA in order to identify individuals who are persistently infected by this class of viruses but are asymptomatic? The memo explained that the viral agent could well be residing dormantly in the chromosomal DNA.[4] One would not expect to find significant quantities of virus in the peripheral blood, making a direct immunological approach to the detection of viral antigens of limited value for complete detection. Since in July 1985 only indirect immunological approaches were available, their understated claim had chilling social implications. If the virus did have a "dormant" period, it provided a powerful public health justification for developing a DNA diagnostic tool, although not necessarily a sufficiently compelling commercial case for investing in one. One area of interest, however, had a potentially wider clinical, hence potential commercial, value. An assay that could detect infected cells without the need to culture virus would be exceedingly helpful to physicians in following patients through the course of therapeutic regimens because it could accurately and rapidly measure therapeutic efficiency.

The first order of business was designing the primers. This task turned out to be a great deal more troublesome than expected. The published sequence data on four retroviral variants (assumed to be variants of the "same strain") suggested that the region of the *gag* gene would be a good target site. The first task was to produce ample target material to experiment on. As in any new system, they encountered a range of technical problems, some quite time consuming to resolve. By January 1986, new primers had been successfully designed to definitively distinguish the AIDS virus from HTLV-1. The primers, as well as the rest of the experimental system, were now working significantly better. Although no single primer pair had yet been designed that could identify all the positive samples, a set of three primer pairs had successfully identified all positive samples.[5]

Interview: Shirley Kwok (continued)

PR   Who trained you in PCR basics?

SK   I don't think anyone was trained. We were given a protocol written by Randy and Steve, and we tried it. It was an exciting period; there was a lot to learn and everyone was contributing. It was a feasibility study. I remember that people were concerned about us working with the HIV specimens, and so to do a PCR we had to set up all the water baths in the P3 biologic containment facility—gowning ourselves up, mask, goggles. It was really uncomfortable to do thirty cycles of amplification and difficult to keep track. "Has it denatured? Is it annealing? Where am I in the cycle?" That went on for several months until me and another guy in the lab told Tom, "If you don't get us a machine, we're quitting."

PR   The first prototypes were beginning to be available?

SK   Yes, soon after. The first instrument was a robotic built in-house by Cetus engineers.

PR   What kind of results were you getting?

SK   They were beautiful. We had a collaborator who sent us our first set of coded AIDS blood samples. We worked up ten specimens. We had one positive out of the ten. I remember clearly the day we broke the code. We were elated, dancing up and down, when we discovered it was the only positive sample of the group. We rapidly became involved in many more studies. But then we were among the first to experience problems with contamination that really bogged us down for a long time. All of a sudden we were getting false positives. I didn't understand what was going on. It was frustrating for me because I was doing a lot of this myself. We'd go to group meetings and I'd have nothing to say but "I've got these positives, but I've also got these false positives." I don't think people at that time appreciated the power of PCR. So, the response I would get would be, "You're just sloppy." There were no special setups to do on the bench.

PR   So the spatial organization of the lab was an important part of the technology. How long did it take to figure out what was going wrong?

SK   Within a couple of months. We were setting up experiments on the bench, and with each experiment we'd set up a bunch of negative controls. Occasionally, a few negatives would come up positive, but over time the problem became progressively worse. Speculating that contamination may be an issue, we'd wipe down the bench, change bench coat, and bleach the pipetters. Although it seemed to have helped, there continued to be a low level of sporadic positives. We went so far as to set up the PCRs in a lab across the street where PCR had never been performed. But the sporadic results continued to haunt us. We didn't know if the contaminants were introduced during PCR setup or during amplification. As a consequence of these battles with contamination, all HIV PCRs were subsequently set up in a "clean" hood dedicated for that purpose. Extraordinary care was taken to insure that the hood and all the equipment used were free of previously amplified product.

We did feasibility studies forever. We ran a large number of drug trials. It wasn't a quantitative assay. I don't think we fully appreciated what we were asking of the assay. We didn't know enough about the PCR technology yet to even evaluate it from that perspective.

PR   Was there a commercial emphasis?

SK   It wasn't so much commercial. I believe Kodak might have come into the picture, and they were working on the technology and formats as well, using HIV as their model. But I never felt any pressure to get a product out. To be honest, the primers we identified early on continue to be candidate primers for the assays. So, it's not as if we weren't satisfied with the performance of the primers. The Kodak situation . . . never really blossomed.

TURBULENCE

During 1987, at the same time these scientific and technical advances were being made at Cetus, the company went through a turbulent period of internal conflict marked by strained relations between Robert Fildes and the scientists heading R&D, especially Jeff Price and Tom White. These conflicts were ostensibly over how best to manage the pharmaceutical operations and strategies of the company: what the project and clinical management structure would be, the extent to which the company's business officers would micromanage, how clinical trials should be handled, and so forth. Players on both sides agreed with Cetus, as an expanding company operating in a changing business, regulatory, and scientific environment, had basic organizational issues to work through. At stake was the company's future, not to mention their own. From the perspective of research and development, the question was whether Cetus was to be managed by a team of individuals who each had experience, responsibility, and authority in his or her own area, or by an autocrat. From the CEO's perspective, the question was, What organizational changes would be necessary to make Cetus a profitable company? In his initial years at Cetus, Fildes worked hard at representing himself as the person who had brought "focus" and "business sense" to a floundering company. This representation was a valuable asset to Cetus in the financial and journalistic worlds but was resented as untrue by key members of R&D at Cetus itself.

Cetus had been run with a high degree of organizational flexibility during its first decade. The advantages of such flexibility were a generally good working environment and a large degree of autonomy for the scientists. The disadvantages were a continuing lack of overall direction that resulted in a dispersal of both financial and human resources and in continuing financial losses. With the company's expansion, the issue of organization came to the fore. There was constant negotiation and eventually open conflict over who would make the decisions and, more specifically, over how to manage a number of simultaneous projects, each interdisciplinary and each proceeding at a different speed.

During the second half of 1986, strains surfaced on several fronts. Very long work weeks, which had been common for many years during the race to develop the interferons and lymphokines, were taking their toll. At first informally, then more formally, discussions began about possible reorganizations. Senior R&D scientists agreed that while they remained committed to a mixed project and departmental structure, it would be helpful to have one person in charge of the overall direction of projects with authority to establish priorities and allocate resources across all departments. They mutually agreed that this person should be Tom White.[6] Jeff Price would remain as the overall head of R&D. This arrangement would free White to focus strictly on the scientific merits, resources, and progress of projects without involvement in departmental or personnel management issues. Because Cetus's business management had simultaneously been lobbying for reorganization and the adoption of a more project-oriented approach, Price anticipated that Fildes would be receptive to reorganizational initiatives, even if he disagreed on specifics. However, in February 1987, when Price presented R&D's reorganizational proposal to Fildes, his reaction was strongly negative. Price remembers Fildes insisting that *he,* Fildes, would take charge of any reorganization at Cetus.

INTERVIEW: ROBERT FILDES

PAUL RABINOW   Why was a reorganization of R&D in order in 1987?

ROBERT FILDES   It was a very difficult time. It was clear to me that we had to become more than a good scientific company; we had to develop the skills to take the science beyond that to products. I began to get frustrated that we seemed to be moving through that phase extremely slowly. My colleagues on the management team were getting frustrated, and I think Jeff Price was frustrated and we talked about it numerous times, and they just didn't seem to come up with a solution. The other dynamic that was going on was with the competition. I was seeing other companies who had started later, who had less resources than we had, making it into the clinic faster than we were. I concluded that what we had to do was create a product management–based organization. That particularly required somebody who was interested in managing those pieces rather than being an expert in any one thing. The analogy would be a nuclear reactor—you wouldn't expect a nuclear physicist to build the reactor. You'd expect him to design the reactor but then you'd go out and hire engineers, plumbers, electricians, bricklayers, and all the rest of it to construct that reactor according to those plans. And you'd pay a general manager to make it happen in the right way so it was done efficiently. What we were facing was we had designed the reactor—products. What we needed was a series of organizational changes to get somebody who could get all the pieces together to the ultimate product.

PR   Jeff and Tom interpret your reactions as an expression of a lack of trust in their judgment and as an attempt to undercut and reduce their authority within Cetus.

RF   That is an overplay. They had four years to show me what they could do. I had worked hard with them and discussed this many times, saying what can we do to improve our performance, never criticizing the research end. Having not been able to work out a solution, something else had to be done. My job is to make the company successful, not any one piece of it. They took it personally. Jeff came back with a counterproposal: "give Tom everything." I said there are two reasons why I can't accept this thing. First, we need more people on this on the management level. You are just going to kill Tom. Number two, Tom is our best scientist, scientific manager. I don't want to take him out of basic research. He's discovering products. I don't think he's ready for the position. They were all very unhappy with the decisions. I basically had a revolution on my hands.

PR   What was Ron Cape's role during this period?

RF   Let's get it straight. Ron Cape had *zero* influence on what was going on at Cetus for the eight years I ran Cetus. And that was part of the agreement. He was a traveling figurehead. He went out and talked to people about biotech, he went looking into how to set up educational systems that would bring the Japanese and American scientific communities closer together. His function at Cetus was to set up board meetings and be there to chair those board meetings. In between, I'd be lucky if I saw him once a month and then I would make a point of spending two or three hours. I would update him about what we were doing in the company. Throughout this whole period he did not pass any comments or views on anything. He stayed out of it. That was our understanding. He stuck to the rules. He would back me at board meetings. He had no input into the running of the company at that point.

In April 1987 Fildes proposed his own plan of action: Judy Blakemore, the IL-2 project manager, would become head of all projects. Fildes considered Blakemore an excellent candidate: "She had the raw qualities I was looking for. On top of that she was an ambitious person with drive, a workaholic; she was incredibly thorough about anything she tackled; why not give her a shot at it?"[7] The plan's cornerstone was that she would report directly to Fildes; henceforth, project management would be under his direction and separated from R&D. Information would flow through Blakemore to him and back again to her and only then to the rest of R&D. Price, White, and their colleagues all felt that Blakemore lacked the experience, expertise, and credentials (she had a business degree and no doctorate in science) to manage all the projects. She would need a long time to "get up to speed." White's most important reservation was that Blakemore would not be able to make independent technical evaluations and judgments of competing projects and would be forced to rely on the "fuzzy" criteria of the project manager's "credibility" or her "trust" in a particular individual. R&D felt that the proposed change would be worse than no change at all. In their eyes, the effect of Fildes's decision was to split R&D, and therefore to weaken it. R&D countered with different reorganization proposals aimed at keeping control of the projects.

This sparring was taking place in the shadow of efforts to raise more money through a second limited partnership offering. Fildes accused the R&D people of sabotaging these efforts, a charge they adamantly denied. They had no interest, they protested, in hurting the company to which their fates were tied. Both sides were mobilizing their allies for a final showdown. Hypothetical vote counts of the board of directors over ousting Fildes became the subject of informal discussion and eventually of formal meetings. The consensus was that although the board was divided, the majority would probably back Fildes.

At the meeting with Price and White, Fildes set forth a proposal that would have effectively made him Cetus's chief scientist. Fildes asked White if he saw any problems with that arrangement. White reacted, in an even tone, with uncharacteristic bluntness, "Well, you are not qualified for the job." The phrase simply came out of his mouth. Price was astounded; he remembers that Fildes looked as though he had been hit by lightning. He turned red and declared something to the effect that he was most certainly qualified for the job. The meeting was quickly adjourned. For the first time, White began seriously thinking about leaving Cetus.

After discussions among the top research scientists, who concurred that such a plan would be catastrophic, Price was delegated to talk to the only person in a position to mediate the crisis, Ron Cape. Cape convened a meeting of the executive officers of Cetus without Fildes to hear R&D's complaints. While agreeing with some of the criticism, they did not agree that the situation was at a breaking point. Nothing was resolved. Another meeting was planned. Cape informed Fildes, who reacted with fury at what he took to be a confirmation of his fears that plots were being hatched to oust him. The scientists felt equally betrayed. As David Gelfand put it, "We thought it was *our* company."[8]

The meeting was held the next day with the same cast and with Fildes present. Cape presented a compromise proposal that would have given White, as a senior vice president, nearly a free hand to organize and run R&D on his own terms, with the proviso that he had to report to Fildes and he had to accept Judy Blakemore as his senior staff person. White and Price remember Cape making the proposal in a tone that approached pleading. The R&D team left the room and decided that the proposal left Fildes's basic agenda unchanged. Price's offer to resign was rejected by the R&D group. They returned to announce that they found the proposal unacceptable. R&D knew that Fildes and his allies would have found it extremely difficult to fire White, Price, and the others, as such action would devastate the company's scientific programs. Even worse, it would look terrible to investors, and Cetus's stock was vulnerable. It was Cape and Fildes's turn to leave the room; on their return, they asked, "What now?" The situation was blocked. Another round of proposals and counterproposals ensued without any progress. White, Price, and others were set to resign.

Price awoke early one morning with a plan: he proposed limiting Blakemore to authority only over projects that were in the clinical phase, as Fildes wanted. Other research projects would continue to be managed by R&D. After negotiations, Fildes agreed. The deal (starting 1 June 1987) included a formal agreement, valid for three years, that senior scientists could not be fired without a one-year severance package (something Fildes had already arranged for himself). R&D had retained control of its activities but, in another sense, Fildes was the winner. He had imposed Blakemore, and the R&D scientists felt that they would still have the double load of their own work and hers.

By January 1988 White had concluded that nothing fundamental had been resolved. He calculated that with the one year's salary he had negotiated as part of the severance arrangement, he could afford to go back to work full time in the lab. Scientifically this was an exhilarating moment; PCR was just beginning to have a major impact on the study of molecular evolution, and White was eager to reinvigorate collaborations with Berkeley colleagues on strategies for using PCR in this area. This collaboration might also reopen the possibility of a university position or government funding. In February White announced his resignation. He remembers both Cape and Fildes expressing disbelief and angry puzzlement over his motives. They proposed a mutual face-saving device that covered over the internal dissension while leaving White's options open—why not call his departure a "sabbatical"? White agreed, and spent 1988 developing PCR methods for molecular studies of the evolutionary biology of fungi.[9]

NEGOTIATIONS: PCR FOR IL-2

Given that PCR had no immediate relevance to cancer therapeutics, it was normal that it be assigned a niche in commercial relationships secondary to the company's main direction. Cetus's decision to look for partners with more market experience and product development expertise in diagnostics and instruments for research meant that PCR would be developed commercially outside by others. In February 1986 Cetus signed agreements with Kodak to develop human *in vitro* diagnostic tests. Cetus had already entered into a joint venture with a major instrument company, Perkin-Elmer, in December 1985. Although the commercial importance of both the reagents and thermocycler market was initially missed—PCR was not mentioned in the initial agreement—it rapidly became apparent that this market was a profitable one. Both companies soon saw PCR as the main focus of PECI (Perkin-Elmer/Cetus Instruments). By November 1987 the first PCR reagent products and the final DNA thermal cycler had been introduced.

The basic PCR patents had been issued in June 1987. During the summer of 1988, management began drafting a PCR business plan. In November 1988, Cetus announced the creation of a new PCR division, to be headed by a businessman. John Sninsky was named director of research and Ellen Daniell director of business development for the division. Diverse and, at times, crosscutting forces led to the decision. Those in favor argued that it would serve as a means to focus research and to identify Cetus's scientists as explicitly working on PCR. At the same time, setting up a division would increase the visibility of PCR to the investment community, to trade journalists, etc. Those who opposed the idea worried that it would add another layer of management, that it would exacerbate the tensions between business and R&D, and that it was a step in the direction of selling the rights to PCR. Ultimately the creation of a division meant that PCR became more "salable."

The potential of PCR was becoming clearer, both within Cetus and in the larger pharmaceutical and instrumentation world. Kodak was beginning to realize that PCR might become a powerful tool and a valuable property. Although the main emphasis and priorities of the three-year program were to produce immunodiagnostic tests for the physician's office, Kodak had supported some PCR product development, mainly on HIV and HLA, and had funded some feasibility research on other tests. Other large companies also were beginning to approach Cetus about potential PCR applications and/or licensing agreements.

As the termination date of the Cetus-Kodak agreement approached, Cetus's management felt confident that it would be able to negotiate more support to PCR. Capital was at a premium because the final push on IL-2 was under way. Cetus began considering negotiating with more than one company to split the commercial rights with Kodak. During 1988 Cetus held discussions with Du Pont, Abbott, and others. At this time, Hoffmann-La Roche had entered the picture as one of the companies interested in the diagnostic rights to PCR when the Cetus-Kodak program came up for renewal. Cetus and Hoffmann-La Roche began a long process of negotiating a joint arrangement in which essentially (although not explicitly, because of legal constraints) Cetus would exchange its diagnostic rights to PCR for the rights to develop and sell IL-2 without infringing Hoffmann-La Roche's license of the patent on the gene.

Since Kodak had been spending approximately $6 million a year for three years in its joint agreement with Cetus, negotiations began in the $10-million range plus continued support for operating expenses. Cetus was strapped for cash, and such sums would be very helpful. Money was not the only variable. Hoffmann-La Roche owned the exclusive rights to the basic patent on recombinant IL-2, through the Taniguchi patent it had sublicensed years earlier from the Japanese company Ajinomoto. Above all else, Fildes wanted the freedom to commercialize IL-2 under this patent. Although in 1988 legal questions were still unresolved, it was likely that Hoffmann-La Roche's patent could give it "blocking rights" on Cetus's IL-2, even though Cetus had developed a variant, its "mutein," on which it held a U.S. patent. In the United States, patents do not depend exclusively on priority of filing the application but give priority of invention to work performed in the United States. Taniguchi's cloning work had been done in Tokyo at the Japan Cancer Research Foundation. Yet Hoffmann-La Roche might be in a position to prevent Cetus from selling the flagship product on which the future of the company was being wagered, at least in Europe and Japan. The legal situation was unclear, but legal battles were costly and time-consuming.

During 1988 Cetus had to decide whether, having lost the race had against that incredibly crude and vulgar approach that the guy had all the time, and the way he would beat on people, and me included. So I felt like my job was going to be impossible to do, the direction of the company was going down a no-win path, and the value of all my equity, which was a substantial part of my personal wealth, was going to diminish rather rapidly. So finally [in March 1990] I told Bob, "Look, it is not going to work. I'm not going to put my shoulder behind the wheel all this time." We negotiated terms of parting; then I left. I was free and clear to do whatever I wanted to do with my stock, which I began unloading as fast as the market would tolerate it. Shortly after that, a number of other people decided to leave. It became a watershed kind of thing.

Bob basically took the role of dictator to the extreme . . . and they rebelled . . . and immediately unleashed all those forces that I'd been holding in check. That led to Bob's downfall—that plus the IL-2 advisory committee.

P R   What was your evaluation of the period 1987 to 1990?

J P   Well, I always felt IL-2 would be okay. I felt the people's expectations of what it would take to get it there were just unrealistic. And that it would come, but it would just take time. What we should have done was focused on IL-2 as the top priority, but we should have had #2 and #3 priorities, and then we should have had #4, #5, and #6 on hold. He said, "To hell with everything else." And that had two effects: One, at a certain period of time, you're betting everything on IL-2, and you're doing it in a premature fashion. Two, you create chaos because of what you did to all the other things.

One of the things that got hurt was the PCR effort. So I think he just couldn't keep all the balls in the air, and panicked and went for one. Which was just the opposite of his behavior earlier, where he wouldn't—in a rational way—let us reduce the number of balls in the air to something manageable. We fought for years, then at the end he collapses it down to one without a plan. Bad judgment. There was no reality check.

The FDA advisory panel's decision to send Cetus's application back for more data was made at the end of July. Within a week

to clone IL-2, it should abandon IL-2 altogether and move on to another protein. Other companies did decide to pursue that strategy; only Amgen and Cetus gambled and continued to pursue their research and development. Although the two companies developed recombinant variants of IL-2 and were soon locked in a legal battle with each other, Cetus had a lead over Amgen and became the major supplier of IL-2 for clinical studies. Cetus's aggressive patent stance and its successful defense in the courts eventually forced Amgen out of the competition. Hoffmann-La Roche, however, was a much stronger competitor than Amgen. At Cetus, the internal debate continued over how much fighting and how much negotiating to do with Roche.

Cetus had in fact approached Roche earlier about entering into partnership, but Roche had declined; IL-2 was not a major product for it and Cetus's mutein didn't interest it greatly. Cetus raised the issue again in 1988, and this time Hoffmann-La Roche was indeed interested, not in Cetus's IL-2 mutein, but in PCR. Having decided that PCR was an extremely promising property—the most powerful DNA probe and amplification technology available—Hoffmann-La Roche also decided that it wanted no three-way partnership with Kodak or anyone else. Two agreements were reached. Under the first, Roche would fund diagnostic research at Cetus for five years at $6 million per year and pay a significant royalty on the sale of jointly developed diagnostic products and services. Roche also purchased warrants for a million shares of Cetus stock at fifteen dollars per share, a figure three dollars above the price on the stock market.[10] In the second agreement, Cetus obtained freedom from suit under Roche's IL-2 patent, and the two companies agreed to share clinical data.

As a deal between Cetus and Hoffmann-La Roche became increasingly likely, Hoffmann-La Roche asked Price for his recommendation of an in-house person who might be available to handle the PCR operations if Hoffmann-La Roche acquired the rights. Price suggested White. White was interviewed in January 1989 and hired in March 1989, one month after Cetus's contract with Kodak expired.[11] White began his work, still housed in a Cetus building, while waiting for the results of the patent trial between Du Pont and Cetus, which would decide the ownership of and commercial control over PCR.

ENDGAME

Increasingly during 1989 and the early months of 1990, there was a real question as to how long Cetus could continue as an independent company, given its continuing financial losses. There were many significant departures among the R&D staff, including those who were handling the IL-2 clinical trials. All the top R&D, clinical, and regulatory managers resigned. In White's view, these departures were directly the result of Fildes's management style, and it was difficult for him to envision a scenario in which new people could be hired to correct the situation. Ron Cape, who was at Cetus only infrequently, had shown that he was not going to take a stand against Fildes. Several former senior scientists at Cetus feel that Cape's passivity was certainly as responsible for the situation at Cetus during this period as Fildes's aggressive style and IL-2 strategy. The potential failure of Cetus posed a problem for Roche: if Cetus were to be acquired by another company, the PCR project could well be put in jeopardy.

During the early summer of 1990, yet another confrontation took place between Fildes and scientists from R&D. David Gelfand organized meetings with Cape and others to tell them that, in his view, Fildes was on a path that would destroy the company. Although they had ignored Gelfand's warnings three years earlier, it was more difficult to ignore them the second time around, after every top R&D manager in the company and other senior scientists had departed. Gelfand and the others were convinced that the company was not going to be rebuilt with Fildes at the helm. Gelfand bitterly remarks that at no point during this crucial period did the board consult Cetus's senior scientists.

INTERVIEW: JEFF PRICE

PAUL RABINOW   How was your relationship with Fildes?

JEFF PRICE   I worked with Bob about eight years; when I left Cetus, I'd been there about fourteen years, so I had pretty much understood what the opportunities and the limitations were of Bob's management of the company. I had considered

leaving several times before; the closest I'd ever come was in 1987.

He may have been correct in criticizing some of what we'd done. I had given him a blueprint that I'd worked on for revi ing things; he basically avoided discussing it with me for wee and weeks and weeks. And then decided on a reorganization plan, nothing like the one I had proposed to him. It was an i portant benchmark, I think, because even before '87 there w fair amount of concern among many of us, about how Bob operating. . . . But until he did that, there wasn't a large am of overt evidence that he was really going to take actions th would be strongly detrimental to the company. The thing have to understand about those actions is that they signale beginning of the end because they said Bob had lost confic in the overall scientific management. He kind of said, "I better than you how to manage this organization. Theref going to take some steps and you're just going to like the lump them." He picked someone who had no meaningfu tific credentials to be a kind of a czar. There was no way cize her because Bob took it personally. During this peri time I was beginning to lose my best people, my best ser tists, my best clinicians. It was very clearly falling apart.

PR   But you weathered that?

JP   We got through it. We decided that we didn't ha port from Ron, we didn't have support from the board took off and a bunch of other people took off with us, thing would collapse. And it was gong to be bad for e Everybody was going to suffer, and the company's op to be successful was going to disappear. So we decide going. We wrote agreements that would cover us. Bi while, all during that period, I had been searching fi to effectively resolve the problems that had been acc and to see if it was possible to resurrect anything ou nizational shifts that had been made.

It finally got to the point where I realized I'd tri try to hold this thing together. It was unrecoverabl not simply my own selfish, personal position, thou had a part in it. After a while I began to lose the i

Fildes was informed by the board that he should resign by 15 August or he would be fired "for cause." Fildes resigned, receiving what some well-placed former Cetus scientists consider to be an extremely generous "golden handshake," including three years' salary, vesting to large amounts of Cetus stock, and continuing remuneration for a consulting role with the company.

INTERVIEW: ROBERT FILDES (CONTINUED)

PAUL RABINOW   The R&D people think that IL-2 was pushed too rapidly. Do you agree?

ROBERT FILDES   I don't think we rushed anything. We started working on IL-2 in 1981. We didn't take it into the clinic until 1984, and got it approved in '90–'91.[12] That's hardly rushing, a ten-year period, which is pretty standard. By industry standards IL-2 had an average track record for a drug. These guys had no prior experience of developing drugs. They were looking at these decisions in purely scientific terms. It's not that simple a decision. Their opinion was wrong. Scientists! A scientist, God bless his socks, always wants to develop the Cadillac. In the real world of products, whether it's medicine or anything else, you can bring products to market that help a situation without necessarily being the ultimate Cadillac. That's true of drugs, of cars, of anything. I'd say, "Come on, guys, let's get a few Fords on the way to the Cadillac. We've got to pay for the Cadillac." IL-2 was in that category.

PR   What happened with the FDA?

RF   We worked real hard, everybody pulled together, made our submission, waited for the FDA to call us before the advisory committee. That took place over [the course of] a year or so. It's a real signal about your chances for getting the drug approved. Big event. We went before the committee and they said, "Look, we are not questioning whether or not this drug is efficacious. It clearly does benefit patients with renal cell carcinoma, but unfortunately it is a difficult drug to use and it has toxic side effects, so it has to be managed carefully. We want to be sure we understand the patient category that it is most beneficial to, because there is a risk aspect to it. We want you to go away and re-examine your data and show us which of the subgroups of patients benefited most from the drug. When you've done that, you can come back and resubmit." That's what really happened.

P R   What was your reaction to that decision?

R F   I was terribly disappointed. I felt that this was a delaying tactic. Obviously it was going to be very costly to the company. On a personal basis, I felt that it was denying patients who would benefit from the drug. The drug was already being used in Europe. I was mad. I was upset. I was concerned. I was frustrated. I expressed some of that. (Laughs)

P R   Do you think you made any tactical errors?

R F   Politically, it was a mistake. To this day I believe what I said was truthful and I was sincere about it, and subsequent events have proved me right. But I guess as the CEO of the company you have to be willing sometimes to bite your tongue and play the game and I didn't. So my judgment on that issue was bad.

P R   Basically this was a minor glitch that could have been ridden out?

R F   Oh, absolutely! I misread the FDA's concern about the toxicity side of the drug. I hadn't calculated that properly, and the reason was I thought about all the chemotherapeutic drugs that had been approved. My call on that was off base. Ron Cape was very embarrassed that I pointed out that the nonapproval of this drug was going to cost lives. It was true. Twenty-four hours later I was back saying, "Okay, guys, what do we have to do to get this thing back on track." I was absolutely confident.

We didn't have any financial worries. We had $150 million of cash, an opportunity if I wanted to sell PCR for $350 million. I had five other products in the clinic coming along. This was someone giving me a black eye along the way. I was bitterly disappointed, obviously, about the stock drop. That's the way the market reacts. Not a surprise.

What surprised the hell out of me was the board of directors. Ron and the board of directors, without talking to me about it, made a decision that we were in a disaster zone. We had to change gears and do something different, look for a big daddy or sell the company. There was no consultation on this or anything. The first I knew Ron was leading the charge; the first I had any inkling at all of it was when Ron announced that this is what was going to happen and I wouldn't fit into the company's plan. Panicky—unnecessary move that led to my departure. We agreed not to go into the details when it happened. I have a very different opinion about the future of this company. One of my biggest mistakes was to not pay attention to who he was putting on the board.

P R   How long did it take you to get over it?

R F   I was mad. This clown is about to ruin the company. What does he know anyway? I had spent twenty-odd years in the industry. Along comes this joker and tries to put a black mark on it. It took me awhile to get over it. I took it very personally. It's like being in a marathon and you run twenty-five miles and then somebody says to you, "Oh, don't bother to finish, just step out." Although Ron stepped aside, he never let go. I will never forgive him for that." [13]

CODA

The *Wall Street Journal* headlined the story on 17 August 1990: "Fildes Quits Top Cetus Job in Wake of FDA's Rebuff." The story continued:

> Cetus Corporation's embattled president and chief executive officer, Robert Fildes, resigned under pressure yesterday, two weeks after a Food and Drug Administration panel rebuffed its cancer drug interleukin 2.
>
> The hot-tempered Mr. Fildes, 52 years old, is being succeeded by Cetus veterans who seem to have been selected to neutralize his combative ways and woo FDA regulators whose help Cetus needs: Ron Cape and Hollings Renton. . . . The company, burdened by losses stemming largely from its $120 million investment in IL-2, also said

it would reduce its work force by 100 from 950. . . . Cetus, its morale at its lowest ebb ever, has been slapped with at least two lawsuits by shareholders angry over the IL-2 debacle. . . . industry analysts cited his [Fildes's] style as a key problem, his stance was characterized as "belligerent." . . . Renton said "we have to rebuild credibility."[14]

On 28 February 1991 Cetus's PCR patents were unanimously upheld in its patent trial with Du Pont. On 23 July 1991 Chiron confidentially agreed to buy Cetus stock for $660 million, for which it would receive approximately $330 million from the sale of PCR and the rest in cash and assets.[15] On 12 December 1991 Hoffmann-La Roche finalized its agreement to buy Cetus's PCR technology for a reported $300 million, and Chiron publicly announced it had acquired Cetus's remaining assets.

CONCLUSION

A Simple Little Thing

The objective person . . . in whom the scientific instinct, after thousands of total and semi-failures, for once blossoms . . . is certainly one of the most precious instruments there are; but he belongs in the hands of one more powerful. He is only an instrument; let us say, he is a mirror—he is no "end in himself." . . . he waits until something comes, and then spreads himself out tenderly lest light footsteps and the quick passage of spirit-like beings should be lost on his plane and skin.
—Friedrich Nietzsche, "We Scholars," *Beyond Good and Evil*

I have chosen to interpret Cetus Corporation as a fortuitous space of experimentation. In that space one may see a certain kind of instrumentalization of sites, subjects, and objects coming together, for a time, into a contingent ensemble. Had I adopted a different perspective—one embodying a different practice—I would have produced a different diagnosis of Cetus and of what happened there. How typical the configuration I have identified is, was, or will be can be debated and contested. I have absolutely no idea how many "Whites" or "Mullises" there are out there, even if one knew how to study such things. My argument does not turn on the numbers, presumably any more than Merton's argument is logically devastated by showing the presence of values in disharmony with scientific norms, or Shapin's would be shaken by documentation of venality among the gentry, or Nietzsche's by a hypothetical psychometric study demonstrating that "spirit-like beings" don't exist.

As a way of completing my project, I solicited a "summing up," an "accounting," from Randall Saiki, Henry Erlich, and Tom White about PCR and the awarding of the Nobel Prize to Kary Mullis.

RANDY SAIKI   I have mixed feelings about it. I feel bad that I couldn't feel better about it. Kary has his good points and bad points; the point is that we worked together. My reservation about it is that it will give him a wider audience for his story, which changes a bit from year to year. It's a fable; it's not really how the science was done at Cetus. It's Kary's twist on things. We've been willing to give Kary the credit for the invention. We don't get the reciprocal treatment from Kary. Had we not been there, he would not have done it on his own. I believe if it had not been for this group being there, nothing would have come of it. He would have dropped it and moved on to something else.

Awarding the prize to Kary alone is not right but [awarding the prize to] Kary and other people is also not right. Our work was good, solid work. I was a technically good grunt in the lab in 1985. We knew what needed to be done. It was kind of difficult. He wanted to do it his way. We wanted to get OR to work. We started to do what needed to be done. Kary was going to take an impossibly long amount of time. Almost like we were competitors in the same building. We did a very good job quickly. We knew what we were doing. We had more experience. We constantly told him to do things. To prove to people that's what it is. It just got worse and worse as we got further along on the project. Perhaps we could have been more sensitive, slowed the project down. But we didn't want to be held back.

He was naive enough not to expect it to work. He knew just enough molecular biology to understand how he might do it but not so much that he would be discouraged from trying.[1]

HENRY ERLICH   On the one hand, I was delighted to see that PCR as a technology was recognized by the Nobel Committee, but I was frustrated that awarding the prize to Mullis alone certified an account which Kary had created which was not true. Mullis had a great idea, which he followed up with years of misrepresentation and self-promotion. Rewriting history was more productive than writing papers.

He never really got it to work. It would have been appropriate to include others on the award. I wished I could have felt happy for him, but I am sufficiently angry at him for his years of misrepresentation that I didn't. On the other hand, it is their prize and they can do whatever they want with it.

In terms of Mullis's PCR creation myth, it was accepted in part because it was never challenged. Many of us who were at Cetus were functionally muzzled. We were told not to upset Kary because they thought it might endanger the patent. He held the cards. The management at Cetus and Roche were concerned that Mullis was unpredictable. There was no assurance he would support the Cetus part of the patent. They had to take great efforts to mollify him, make sure that no one at Cetus was going to do anything that was going to offend him. So as a result, his assertions have not been challenged or refuted by any of the people involved.

I don't think Kary initially understood the real significance of PCR. For Kary, PCR was largely a synthetic method that would allow one to synthesize large amounts of something; he was, after all, a synthetic chemist. The part that my lab contributed with Randy and Steve was the analytic applications of PCR, the amplification from human genomic DNA of a specific gene and typing it or distinguishing the alleles, sequencing. From my perspective, the real power of PCR is as an analytic tool to look at genetic variation. It is somewhat misleading to say it was first applied to sickle-cell anemia. In fact, it was developed using sickle-cell anemia as a model. Then, after beta globin, it was applied to the HLA system.

Committees and science journalists like the idea of associating a unique idea with a unique person, the lone genius. PCR is, in fact, one of the classic examples of teamwork. Many people contributed: the people in my lab, various engineers, Gelfand's group, Sninsky, White. If Kary had acknowledged those people, it would be easier to take. We viewed PCR as a means—a powerful method that allowed us and others to do innovative research. Kary seems to have viewed PCR as a means to celebrity. He started revising history in his first manuscript. In the first version, he thanks Tom for getting out of his way and letting him figure it out. That was written after we had already figured out how to make PCR work. The irony was that the reason PCR was working and that Kary had a paper to write at all was that Tom had asked me and Norm and the group to get involved and see if this idea had some potential. Tom had not stayed out of his way. If he hadn't intervened, there would have been no PCR.[2]

TOM WHITE   Six points: One, this is the first Nobel Prize based on science done at a start-up biotechnology company, which I think is *very* significant. It affirms that creative science was being done in these companies in the 1980s, as creative as that being done in any other labs in the world. Two, the work was rapidly published. The work was published as soon as experimental data worthy of publication were obtained. The decision was made not to keep it a trade secret, even though the inventor himself thought it should be kept a secret. Three, it shows that although they are usually considered to be incompatible, such science and its patenting don't inherently have to conflict. Four, the idea was clearly Mullis's, no one contests that. Five, the key experiments to prove the concept worked were done by a team, not by Mullis alone, in a specific environment. Without that team and that environment there would have been no PCR at that time. Six, there were at the time good research reasons to doubt that the concept would work and to doubt his ability to make it work.[3]

Randy Saiki finds the way credit is distributed a tedious game played by those with too much time on their hands. He has many other things to do. Saiki is irked by what he sees as Mullis's egotism and lack of seriousness, his refusal to acknowledge that teamwork and professionalism made PCR work. Seen from the lab bench, science is a question of demonstrable results, the product of collective activity. If, for Saiki, Kary's "twist" distorts the truth by creating a fable of the individual genius, for Henry Erlich what he sees as Mullis's self-conscious fabulations betray science as an institution. For Erlich, science is analysis; science makes models; the "significance" of science is found in the innovative research its practitioners produce. Erlich's chafing stems from his estimation that the misrepresentation of PCR's importance led not only to a misapportionment of individual credit but to a breach of faith with science. He sees not only perfidy feted as genius but a technical idea mistaken for scientific understanding—these twin distortions vex him. Erlich insists vehemently that my characterization of him in an earlier draft as "bitter" was simply wrong: he is not bitter about his life choices and their outcomes, only "pissed off" about Mullis. Science continues to be highly rewarding for him; he is pleased to have been part of the development of PCR. He has no regrets; he only wants the real story to be told. Erlich wants to write the real scientists back into their appropriate place, in the middle of serious things. Both Saiki and Erlich believe that Mullis distorted—perverted—the real "significance" of PCR.

Tom White is neither fed up nor bitter about PCR or anything else, for that matter. His memo-like remarks personify his disposition: a restrained exuberance over what has been achieved, recognized, and legitimated. White sees no betrayal of his scientific vocation; he is proud to have contributed to a major event, the birth of a momentous invention. He tells no corruption story of irremediably tainted modern institutions; rather, he lauds the creation of a new institutional form, one guided by standard Mertonian scientific norms, in which collective action maximized individual contributions while structurally compensating for individual shortcomings, precisely as they were supposed to do—at least for the required time. White et al. harnessed the virtues and vices of a cast of strong-willed players long enough to demonstrate credible scientific results and the existence and potential of a very consequential scientific technique. If that meant redirecting resources, or pitting the efforts of two teams against each other while regularly patching relations, or leaving Cetus over what he saw as its misdirection, so be it.

If White had been in the audience when Weber delivered his stern peroration on the modernity of science, he would not have had to squirm. Weber proclaimed:

> Science today is a "vocation" organized in special disciplines in the service of self-clarification and knowledge of interrelated facts. It is not the gift of grace of seers and prophets dispensing sacred values and revelations, nor does it partake of the contemplation of sages and philosophers about the meaning of the universe. This, to be sure, is the inescapable condition of our historical situation.[4]

Today, the lure of seers and prophets is faint, at least for those like White for whom the challenge of self-clarification consists in avoiding the snares of credit (both material and symbolic) as a primary goal of one's conduct or even as a post hoc recompense. White's goal, after all, was to contribute to a pragmatic mastery of life's matter, honestly achieved, while making a living doing it. Benjamin Franklin would have approved. When Hoffmann-La Roche purchased the rights to PCR for $300 million, the company asked White to manage its acquisition for it. He agreed, upon condition that he could continue to practice science as he had been practicing it. Tom White was someone you could trust. Everybody trusted him: Mullis, Erlich, Kwok, Fildes, Hoffmann-La Roche. I trusted him. Above all, White trusted himself. Today, Mullis occasionally calls him, Fildes has respectful words for him, Erlich and Kwok work for him, Hoffmann-La Roche pays him.

His "peculiar ethic," to use Weber's phrase about Franklin, is "the essence of the matter. It is not mere business [or scientific or managerial] astuteness, that sort of thing is common enough, it is an ethos . . . an ethically coloured maxim for the conduct of life. *This* is the quality that interests us."[5] In my representation, the curious and distinctive quality in White's work of life regulation is his disposition as the "objective" person, the finely attuned instrument, who reacted and responded to an ensemble of techniques, concepts, experimental systems, spaces, organizational conflicts, scientific currents, and a multitude of conflicting temperaments in a timely and efficient manner.

In the accounts they provide of themselves, these Cetus scientists come across as earnest, intelligent, hardworking, and reasonably optimistic. These Americans are professionals. They present themselves as basically at home in a dynamic environment, one they took an active role in shaping and managing. Their scientific practice, they firmly believe, contributes to a general broadening of scientific understanding and technical mastery, to an eventual improvement of public health, and even to the betterment of society.[6] They are proud to have made something valuable.

It would be easy—but precipitous—to echo Max Weber's brusque dismissal of such a faith in the worldly benefits of science: "After Nietzsche's devastating criticism of those 'last men' who 'invented happiness,' I may leave aside altogether the naive optimism in which science—that is, the technique of mastering life which rests upon science—has been celebrated as the way to happiness. Who believes this? aside from a few big children in university chairs or editorial offices." It would be precipitous because the Cetus scientists share Weber's contempt (they express it as irritation) for what they see as the increasing tabloidization of science in its leading journals or the extravagant claims of the spokesmen for its megaprojects. It would also be precipitous because health is arguably the dominant value of the contemporary world, and biomedical science is considered to be a major contributor to improving health.

Nonetheless, the continuation of Weber's remarks remains pertinent:

> Under these internal presuppositions, what is the meaning of science as a vocation? . . . Tolstoi has given the simplest answer, with the words: "Science is meaningless because it gives no answer to our question, the only important question for us: 'What shall we do and how shall we live?'" That science does not give an answer to this question is indisputable. The only answer that remains is the sense in which science gives "no" answer, and whether or not science might yet be of some use to the one who puts the question correctly.[7]

Three-quarters of a century later, while Weber's diagnosis of modernity remains remarkably contemporary, the sense in which science gives "no" answer has left open a cultural space in which even the question has lost its urgency. Weber's somber tone is both current and archaic; the Cetus scientists do not expect their work to answer such questions. Other objects, other subjects now occupy something like the same space, and therefore other questions need to be posed to them, by them—and about them.

(Re)enter the pragmatic Dewey, striding energetically and confidently past the old European spirits:

> Science is the instrument of increasing our technique in attaining results already known and cherished. More important yet, it is the method of emancipating us from enslavement to customary ends, the ends established in the

> past. [T]he man of science, must bear in mind that practi-
> cal application—that is, experiment—is a condition of his
> own calling.... Consequently, in order that he keep his
> own balance, it is needed that his findings be everywhere
> applied. The more their application is confined within his
> own special calling, the less meaning do the conceptions
> possess, and the more exposed they are to error.... As
> long as the specialist hugs his own results they are vague
> in meaning and unsafe in content.[8]

For Dewey, there is no metaphysical crisis; the loss of the quest
for "true meaning" is actually an emancipation. Clinging to the
chimeras of "the true story," hugging too hard, unbalances one.
Dewey would urge Weber to stop worrying and start experimenting: "That individuals in every branch of human endeavor should
be experimentalists engaged in testing the findings of the theorist
is the sole final guaranty for the sanity of the theorist." Reluctantly,
Weber might almost agree and consent—almost. Speaking to his
German audience about the American students' equation of democracy and science as both individualist and achievement-
oriented, Weber observed that "formulated in this manner, we
should reject this. But the question is whether there is not a grain
of salt contained in this feeling." We might counsel Dewey to consider that, while metaphysics is probably over and while theory
may plausibly be cast as practice, nonetheless experiments and experience and troubles go together. This triad does not always lead
to salutary results, or ones "correctable" by pragmatic action and
further experimentation. The sanity of the theorist, after all, is only
one concern.

Today one might well suggest to Weber that the grain of salt is
surely not the "individual" or "democracy" or "science" or "experiments" but, rather, the power to make things—in this instance,
biotechnological ones. We should no doubt be curious and careful
about what questions we should be asking about them. We should
wonder about who the "we" is. What form of life, after all, is based
in part on tinkering with genetic material and the emergence of
unprecedented events? What form of "life regulation" could ac-
company it? Do "experience" and "experiment" depend too much

on an older notion of "choice," hence on an archaic sense of "risk" and security?

EVENT

PCR is frequently characterized as a "revolution," as it was by a scientist quoted in a 1990 issue of *Science:* "I know the term is overused, but this is a revolution."[9] He then goes on to describe the spread of PCR into different domains of biology and its immense power to facilitate previously time-consuming tasks. The word is not only overused, it is used inappropriately. The article's misleading title, "Democratizing the DNA Sequence," confounds the confusion. Strictly speaking, the title should have read, "Expanding access to clones and genes, by making their transfer between labs unnecessary once one knows their DNA sequence and thus can obtain them by PCR from any human's total DNA."

Mullis, with his flair for parody, mocks the view that PCR is either a political or a scientific revolution. He writes:

> I know of two kinds of revolution in molecular biology. There is the kind where a band of angry, young, well-armed molecular biologists, having fomented their plans in the chill, rarefied air of the UCLA winter symposia . . . converge in the Spring on Bethesda, assault rifle and ugly unpatriotic slides in hand, to settle once and for all the issue of NIH post-doc stipends.
>
> Then there is the other kind, referred to as a paradigm shift, or a retreat to the drawing board, when disappointing data can no longer be hidden or explained by old notions. New concepts become fashionable and new paragraphs have to be written for introductions to papers and grants. Usually there are a number of powerful elders in important places that have to retire or die before things get rolling. . . .
>
> But I do not recall either kind of revolution here on account of PCR. . . . It was just business as usual exploring genes. Things went faster and easier and the range of pos-

sibility expanded. Nobody had to die for PCR to be accepted. It was just a new tool. . . .
Being a simple little thing PCR tends to work its way into many studies. Everyone thinks of their own little twist to put on it to make it work for their own particular problem.[10]

Mullis is basically correct in arguing that PCR is not a political revolution. Although the large-scale structural changes in funding, institutions, and status had many consequences in reshaping the sites in which molecular biology is done and to an extent the type of the scientists who do it, the bulk of these changes preceded PCR and were, in many ways, instrumental in establishing the milieu in which PCR emerged. Mullis is basically correct in stating that PCR is not a scientific revolution. PCR did not emerge as a solution to a growing set of theoretical anomalies in a scientific discipline. Mullis invented a concept, but the historical distinctiveness of PCR lies less in theoretical advances that it has facilitated (as important as they are in their own right) than in the *practice* that accompanied it.

Mullis is quite wrong, however, when he says it was "business as usual exploring genes." First, the business of genes was new. Second, Mullis himself was not exploring genes, he was manipulating DNA. Third, it took time to see that PCR was the target to be amplified. Fourth, seeing, doing, managing required diverse skills and diverse resources. Fifth, his stylized modesty—"just a new tool"—is unconvincing; Mullis is a gifted twister. Sixth, PCR is a distinctive kind of tool, a biotechnological one.

The conceptual, technical, experimental, and managerial "tinkering" that resulted in PCR can be seen as *bricolage,* a term given theoretical import by the French anthropologist Claude Lévi-Strauss in his classic study of *The Savage Mind:* "In its old sense the verb 'bricoler' applied to ball games and billiards, to hunting, shooting, riding. It was however always used with reference to some extraneous movement (*un mouvement incident*): a ball rebounding, a dog straying or a horse swerving from its direct course to avoid an obstacle."[11] Though it is usually the "wire and chewing gum" patching together dimension of *bricolage* that is emphasized, for PCR the *"mouvement incident,"* the swerve, is equally pertinent.

Mullis was a player in a game that was already under way. His rebound from a blocked course of action turned into a swerve and eventually—a potential movement. White's straying from the rules of the game produced an incidental motion as well. These deviations made something new happen.

Within a very short span of time some curious and wonderful reversals, orthogonal movements, began happening: the concept itself became an experimental system; the experimental system became a technique; the techniques became concepts. These rapidly developing variations and mutually referential changes of level were integrated into a research milieu, first at Cetus, then in other places, then, soon, in very many other places. These places began to resemble each other because people were building them to do so, but were often not identical.[12] Thousands of scientists and technicians around the globe began using PCR, multiplying the modifications and feedback—nested PCR, inverse PCR, single-molecule amplification, universal primers, direct DNA sequencing, multiplex amplifications, quantitation, single-gamete genotyping, dUTP/UDG, combinatorial libraries, aptamers, isothermal amplification, sequence-tagged sites, ancient DNA, *in situ* PCR, single enzyme Rt-PCR, long PCR, etc., etc., etc. Learning and making and remaking: new variants of the instruments, practices, spaces, discourses. PCR is more than any of its specific uses—it has the distinctive quality of continuing to produce events. Mullis's original stress on decontextualization has been transformed into a multitude of recontextualizations that then, it is discovered, contain the potential for further decontextualization, further invention.

A simple little thing.

Kary Mullis

Tom White

Henry Erlich

Steven Scharf

Bez Khozrovi, Jeff Price, and Tom White

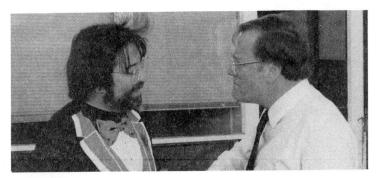

David Gelfand and Robert Fildes

Randall Saiki

Shirley Kwok

Ellen Daniell

Norman Arnheim

A Note on the Interviews

All interviews were conducted by the author in California. The interviewees and the author reviewed the transcriptions and made minor revisions to enhance readability. David H. Gelfand (chapter 1) was interviewed in August 1992 in Berkeley and Oakland; Robert Fildes (chapters 2 and 5) in October and December 1993 in Berkeley; Tom White (chapter 2) in August 1992 in Berkeley; Ellen Daniell (chapter 3) in May 1995 in Berkeley; Shirley Kwok (chapter 5) in January 1995 in Alameda; and Jeff Price (chapter 5) in October 1993 in Richmond. (See the preceding two pages for photographs.)

Acknowledgments

This book could not have been written without the active participation and extraordinary generosity of many scientists. I owe a profound debt of gratitude to Tom White, David Gelfand, Henry Erlich, Steven Scharf, Randall Saiki, Norman Arnheim, Corey Levenson, Ellen Daniell, Shirley Kwok, Jeff Price, Kary Mullis, Christian Orego, Paul Billings, Barbara Bowman, John Sninsky, Fred Faloona, Vincent Sarich, Robert Fildes, Judy Weiss.

It could also not have been written without the instruction and support of colleagues in the social studies of science. Those who read the manuscript and did their best to improve it include Sharon Traweek, Deborah Heath, Ilyana Lowy, David Hess, Gary Downey, Michael Fischer, Soren Germer, Frank Rothschild, Raymondus Krondratas. My gratitude and apologies to the great number of scholars in the social and historical studies of science and technology from whom I have learned a great deal. I regret that it is inappropriate to include more explicit citations to the lively debates of these fields; keen and tolerant readers will find traces abound. I trust that my colleagues will realize that this book seeks a somewhat broader audience, including some who are far less tolerant of the technical language of science studies.

Unusually detailed and perceptive readings were generously offered by Stephen Hall, Hans-Jorg Rheinberger, Michael Panisitti, Adriana Petryna, Bob Laird. They helped to shape the book, and I thank them.

Two friends, James Faubion and Joao Guilherme Biehl, have practiced and exemplified what for me is the highest of virtues, both ethically and intellectually, *philia*. Without their care, this book and my life would have been impoverished.

Mary Murrell at Princeton University Press and Kirk Jensen at Oxford University Press both read a very early draft of this book and commented on it with intelligence and professionalism. Their honesty was instrumental in making the path ahead clearer. Having the good fortune to be in exchange with these editors was a privilege I value deeply.

Susan Abrams of the University of Chicago Press provided unstintingly honest and strikingly productive criticism and support, the kind one can trust. Fortune smiled on me—how else to characterize this encounter with someone who so completely embodies and exemplifies an honorable vocation. *Va bene*.

Although the ethnographic work and the writing of this book is unredeemed by any Federal grant support (long live peer review, may I live to experience it), Suzanne Huttner and the UC Systemwide Biotechnology Initiative were generous in providing funds for computers, for which I am grateful.

To Marc for his constant change and Marilyn for her constancy.

Notes

Introduction

1. I started this project with an inquiry into the Human Genome Initiative, specifically at the Lawrence Berkeley Laboratory, and continued it in a biotechnology company after political struggles over future directions at LBL made it a difficult place to work. During 1995 I began a study of the major French genome mapping center, the Centre d'Etude du Polymorphisme Humain, in Paris.

2. King and Stansfield 1990, 247.

3. Daniel E. Koshland Jr., "Perspective," *Science,* 22 December 1989, 1541.

4. Guyer and Koshland 1989, 1543.

5. Patricia A. Morgan, managing editor, *Science,* letter to Kary B. Mullis, 12 March 1986.

6. Kary Mullis, interview for the Smithsonian Museum by Raymond Kondratas, San Diego, Calif., 11 May 1992, as part of an archival project on the history of biotechnology. Emphasis is mine. I have drawn quotes from Mullis from published or public documents. I have done so essentially for two reasons: first, by the 1990s Mullis was telling the same stories with the same details regardless of context; second, this procedure simplified the legal picture. I have interviewed Mullis a number of times; he has lectured in my Berkeley class on the Anthropology of Science; I played an active role in framing questions and filling in the background for a part of the Smithsonian interview. I would like to thank Ray Kondratas formally for his generosity and openness.

7. Stephen Scharf, personal communication.

8. Henry Erlich, personal communication.

9. Jacob 1988, 234.

10. Norman Arnheim, personal communication.

11. Although he refused to testify at the trial, Khorana later let it be known that he agreed with Du Pont's claim. During the trial, another Nobel

Prize laureate, Arthur Kornberg, supported the claim that PCR was obvious from his own previous work on DNA polymerase. Another Nobel Prize laureate, Hamilton Smith, testified for Cetus against accepting this position.

12. Nor did Arthur Kornberg mention PCR in the first edition of his textbook on DNA amplification.

13. Henry Erlich, personal communication, 10 April 1993.

14. Mukerji 1989, 197.

15. Wright 1986a, 356.

16. Keller 1985; Keller 1992. See also Kay 1993. Many of these themes were first explored by Yoxen (1982).

17. Pauly 1987.

18. For an example see Swann 1988.

19. *Rockefeller's Medicine Men;* Kohler 1976.

20. Dickson 1988.

21. Merton 1973, 267.

22. Mulkay, "Sociology," 51.

23. Collins 1975.

24. Shapin 1994, 410.

25. Ibid., 413.

26. Ibid., 412.

27. Ibid., 414–15.

28. Weber, 1946, 135.

29. Dewey 1953, introduction, 73–74.

Chapter One

1. Rabinow 1992.

2. Office of Technology Assessment (OTA) 1988, 49.

3. Kloppenburg 1990.

4. Eisenberg 1987, 186.

5. The changing conceptions of nature are the heart of the matter. See, for example, Thomas 1983.

6. OTA 1988, 50.

7. Jameson 1991.

8. OTA 1988, 7.

9. Eisenberg 1987, 186.

10. Teitelman 1989, 14.

11. Smith 1990.

12. Teitelman 1989, 17.

13. Dickson 1988, ix.

14. Ibid., 21.

15. Krimsky 1982; Wright 1986b.

16. Comparison of the regulatory environment in other countries highlights this contingency. Wright 1994.

17. Krimsky 1982, 10.
18. Kenney 1986, 27.
19. P. Schaffer, "The architect's strategic role in planning and designing facilities for biotechnology," *Genetic Engineering News* 3 (March/April 1983): 23. Quoted in Kenney 1986, 180.
20. Kenney 1986, 182.
21. Ibid., 179.
22. Wright 1986a, 347. Robert Teitelman's *Gene Dreams* presents several detailed examples of the strategic involvement of several pharmaceutical giants in the fields of biotechnology. Basic information is contained in OTA 1984a,b. An excellent analysis is found in Kenney 1986.
23. Kenney 1986, 91.
24. Ibid., 18.
25. Kornberg 1989, 294.
26. Paul Billings, personal communication, 16 June 1992. Billings, an active spokesman in the genome and ethics world, never heard of any distinguished scientist refusing monetary rewards for serving as an advisor.
27. Kornberg 1989, 289.
28. Ibid., 291.
29. Former Cetus scientist, 8 June 1992.
30. Cetus "spun off" two subsidiaries, Cetus Palo Alto and Cetus Immune Corporation, that were set up and directed by faculty from Stanford University. There were exceptions to this general picture. Nobel Prize winner Joshua Lederberg was a long-term member of Cetus's SAB. He neither placed his students in the company nor took money for his lab. He took his mandate to "think five years ahead," removed from the day-to-day aspects of the labs, quite seriously. Cetus scientists report that Lederberg was neither taken in by the politics, nor involved in putting down the founders, nor particularly hierarchical in his dealings with younger scientists. On the other hand, he apparently often said nothing at advisory board meetings.
31. Cetus Corporation, *Annual Report,* 1981, 6.
32. For example, Cetus's Stanford consultants strongly advocated working on lymphokines and immunotoxins and were opposed to working on any of the biopeptides. Although claims were made about immediate breakthrough results, the immunotoxins work continued for many years without the promised breakthroughs.
33. Hall 1987.
34. Other examples were interleukins 1 and 3; tumor necrosis factor (TNF); lymphotoxin; colony stimulating factors for macrophages, granulocytes, both of these, and megakaryocytes (M, G, GM, mega); and erythropoietin, all of which affect blood-cell production and function.
35. Cetus Corporation, *Annual report,* 1981, 2.
36. Ibid., 1–2.

Chapter Two

1. Kenney 1986, 157.

2. "Cetus, a Genetic Engineering Firm, Plans Initial Public Offer of 5.2 Million Shares," *Wall Street Journal,* 14 January 1981.

3. Ibid.

4. Cetus Corporation, *Annual Report,* 1981, 18.

5. Complementary DNA, or cDNA, is produced from an RNA template by a special polymerase (reverse transcriptase); it may differ from the original gene by having its noncoding "introns" spliced out. A cDNA library, therefore, is a collection of all the various messenger RNA molecules produced by a specific type of cell of a given species, spliced into a cloning vector. Since not all genes are active in all cells, a cDNA library is usually smaller than a genomic library, which can greatly facilitate screening. King and Stansfield 1990, 49.

6. Cetus Corporation, *Annual Report,* 1982, 2.

7. Jeff Price and Tom White, interviews by author, summer 1993.

8. Tom White, memo, 1 April 1981.

9. Ibid.

10. Tom White, memo, 15 December 1981.

11. Cetus Corporation, *Annual Report,* 1982, 2.

12. Ibid.

13. Tom White, memo to Jeff Price, 20 January 1984, p. 3.

14. Interleukins are defined as "proteins secreted by mononuclear white blood cells that induce the growth and differentiation of lymphocytes.... Interleukin 2 is secreted mainly by helper T lymphocytes following stimulation by interleukin and binding of antigen to the T cell receptor. The binding of IL-2 on T lymphocytes causes them to proliferate and to secrete lymphokines"; King and Stansfield 1990, 165. Lymphokines are defined as "a heterogeneous group of glycoproteins (molecular weights 10,000 to 200,000) released from T lymphocytes after contact with a cognate antigen. Lymphokines affect other cells of the host rather than reacting directly with antigens"; ibid., 183.

15. Teitelman 1989, 28.

16. For an overview, see Panem 1984.

17. By Dr. Tadatsugu Taniguchi.

18. Teitelman 1989, 35.

19. Rosenberg 1992, 114.

20. Rosenberg 1992, 152.

21. Shilts 1987, 366.

22. Jeff Price, interview, 5 October 1993.

23. *Kenney* 1986, 171.

24. *Wall Street Journal,* 2 June 1982. Socal had invested nearly $8 million in the fructose project since 1979.

25. "Cetus Corp. Trims Variety of Products to Commercial Core," *Wall Street Journal,* 23 July 1982.

26. Cetus Corporation, *Annual Report,* 1983, 3.

27. Ibid., 2.

28. Seventy-five million dollars was raised from individual investors to build a diagnostic and therapeutics human health-care business, Cetus Health Care Limited Partnership. The company lost more than $5 million.

29. *Wall Street Journal,* 10 May 1983.

30. Rosenberg 1992, 169.

31. Ibid., 196.

Chapter Three

1. Henry Erlich, interview by author, 9 August 1993.

2. Randall Saiki, interview by author, 10 June 1992.

3. Holtzman 1989, 64.

4. Conner et al. 1983.

5. Saiki, interview.

6. Kary Mullis, interview by Raymond Kondratas. All quotations in this section are from this interview unless otherwise noted.

7. Ibid.

8. Ibid.

9. Tom White, personal communication, 15 August 1992.

10. *United States District Court, Northern District of California, E. I. Du Pont De Nemours & Co. v Cetus Corporation,* E. I. Du Pont De Nemours & Co. transcript, p. 1819. 5 February 1991.

11. Ibid., p. 1840 (6 February 1991). Norm Arnheim disagrees, claiming that this was well known at the time.

12. Ibid., p. 1817. According to Corey Levenson, the first machine at Cetus was Canadian but didn't work.

13. Mullis, interview.

14. *Du Pont v Cetus,* transcript, p. 1856.

15. Corey Levenson, interview by author, 20 January 1995.

16. On explanations of discovery in science, see Root-Bernstein 1989.

17. *Sunset Western Garden Book,* 17th ed. (Menlo Park, CA: Lane Publishing Co., 1976), 175.

18. Mullis, 1990, 57.

19. Arthur Kornberg 1989, chapter 5, "The Synthesis of DNA."

20. Mullis 1990, 60.

21. Ibid., 61. This phrasing is not quite accurate. Mullis means that since the template was double-stranded, half of the extended oligonucleotides would have the same base sequences (and half the complementary sequences).

22. *Du Pont v Cetus,* transcript, pp. 1870–71.

23. Ibid., p. 1874.
24. Ibid., p. 1875.
25. Ibid., Mullis deposition, 1 May 1990.
26. Levenson, interview.
27. *Du Pont v Cetus,* transcript, p. 1971.
28. Ibid., p. 1974.
29. Ibid., p. 1879.
30. Ibid., Mullis deposition, p. 60.
31. Ibid., p. 64.
32. Ibid., pp. 74–75.
33. *Du Pont v Cetus,* transcript, pp. 1880–81.
34. King and Stansfield 1990, p. 243.
35. *Du Pont v Cetus,* Mullis deposition, p. 91.
36. Ibid.
37. *Du Pont v Cetus,* Mullis deposition, p. 1883.
38. Ibid., p. 1886.
39. Ibid., p. 1991.
40. Ibid., Mullis deposition, pp. 105–11.
41. Ibid., p. 115. This is the last experiment noted in notebook 1,000.
42. Kary Mullis, "Outline of Accomplishments 6–83 until 6–84," received by Tom White on 8 June 1984.
43. *Du Pont v Cetus,* transcript, pp. 1893–94.
44. Tom White, interviews by author, July 1994.
45. Ibid. According to White, "Arnheim also made very useful comments on the utility of the method; that is, rather than using a plasmid as template (that method might be workable but it would be only so interesting if you were working on plasmids) to really try to do it on a single copy gene—beta globin in human DNA—that is, something like the diagnosis or detection of the sickle-cell mutation, that was really Arnheim's crucial contribution by stating that as a criterion in the method to me. And Kary's always negated that and said, 'Well, anybody could have said that. It's totally simplistic, etc.' But at the time it was really an important criterion."
46. Ibid.
47. Levenson, interview.

Chapter Four

1. *Du Pont v Cetus,* transcript, p. 2600 (13 February 1991).
2. Arnheim 1983.
3. This section is largely based on an extensive interview by the author with Arnheim and Tom White on 5 August 1992.
4. The general strategy for improving diagnostics in the early 1980s turned on increasing the specificity and sensitivity of the probe. Logically,

this approach flowed from the state of the current available technology and attempts to improve it. Its limitations were found there as well. Specifically, Bruce Wallace had recently developed the technique of oligo hybridization on Southern blots, the so-called allele-specific oligo probe. This depended on introducing a large quantity of radioactive atoms within the probe so as to yield a stronger signal. The more radioactivity used, the stronger the signal. The problem was that the high level of radioactivity of the procedure caused the probe to break up. In general the procedure was messy as well as highly radioactive. Thus, although the concept of getting a specific probe to anneal to a specific bound target seemed sound, the available means of achieving the hybridization were unsatisfactory.

5. Arnheim and White, interview.

6. Stephen Scharf, interview by author, 3 August 1992.

7. Ibid.

8. Ibid.

9. Ibid.

10. The patent story is a complicated one. There was a good deal of conflict over its wording and how it attributed credit for PCR.

11. Daniell 1994, 424.

12. White, interview.

13. Ibid.

14. Arnheim, interview.

15. White, interview.

16. Ibid.

17. Saiki et al. 1985, 1350.

18. Ibid. Note 12 in the article refers to an article by Mullis and Faloona as "in preparation."

19. Saiki et al. 1985, 1354.

20. Watson and Crick 1953, 738.

21. Kary Mullis, interview by Raymond Kondratas. He intended to end the paper with an "experiment that demonstrated that you could amplify off the beta globin gene. It would be a single band on a gel. You could either cut it with an enzyme that cut at the wild type or you could cut it with an enzyme that cut something else on it, and you could show that if it cut at a certain place, then it was the wild type; and if you didn't it was the sickle type. It was a nice little nonradioactive diagnostic that took advantage of the fact that PCR could amplify the hell out of a sequence and could make a discrete band on a gel out of it. Arnheim et al. stuck to the old way of doing it. They wanted a publication."

22. Mullis, interview.

23. Ibid.

24. Mullis et al. 1986.

25. Ibid., 263, 268.

26. Thanks especially to David Gelfand for his patience in explaining polymerases to me. On the three polymerases see Kornberg 1989, chap. 7, "Astonishing Machines of Replication," 207–39; Mullis, interview.

27. Because the Klenow fragment, the polymerase used to extend the primers, was "thermolabile" (it was deactivated by the 95°C temperatures necessary for denaturing DNA), it had to be added after each denaturing cycle. Cetus had developed a multiplate liquid handler, Pro/Pette.™ "It was modified to use two temperature controlled aluminum blocks, each with a 48 sample capacity. The front block held the samples (in uncapped microcentrifuge tubes) and was connected, via a switching valve, to two water baths, one at 94°C, the other at 37°C. The back block held the solutions of Klenow fragment, also in uncapped microcentrifuge tubes and placed in the same configuration as the samples in the front block. To ensure the stability of the enzyme solutions, the back block was connected to a water bath set at 4°C. A controller kept track of the incubation times at both high and low temperatures, actuated the switching valve in order to change the front block temperature and prompted the multi-channel head to pick up fresh tips, withdraw aliquots of enzyme solution from the tubes in the back block and deliver them into the corresponding samples in the front block." Oste 1989, 24. This instrument is now exhibited in the Smithsonian Museum in Washington, D.C.

28. It was expensive. Even though Cetus was receiving a substantial discount in price on the large quantities of the enzyme it was buying, the price in 1985 was still close to a dollar per cycle. Hence, a thirty-cycle PCR would cost roughly thirty dollars for the enzyme alone. In 1993 the cost for the *Taq* enzyme was closer to fifty cents for thirty cycles or more. Regardless of cost, the first goal, of course, had been to make the PCR process work; Klenow did that.

29. David Gelfand, interview by author, 2 July 1993; Chien, Edgar, and Treia 1976; Kaleidin, Siyusarenko, and Gorodetski 1980. Why some polymerases are heat-resistant is basically unknown and is probably a function of the specific spatial arrangements of their amino acids. For Cetus, the most troubling aspect of the available knowledge was that the DNA polymerase activity was not strictly dependent on the presence of all four deoxynucleotide-triphosphate substrates. This type of misincorporation was clearly not the way the polymerase was supposed to function in the cell. It wasn't clear why the misincorporation was taking place. One possibility was simply that outside their cellular milieu these enzymes simply did not perform their copying functions very faithfully. The situation needed to be clarified.

30. Gelfand, interview by author, 29 July 1993.

31. Mullis, interview.

32. Gelfand, interview.

33. Ibid.

34. Mullis, interview.

35. Gelfand, interview.

36. Mullis, interview.

37. Since leaving Cetus, Mullis has worked as a consultant for a variety of biotechnology companies. His vita indicates that, even before his Nobel Prize, he was in ever increasing demand as a lecturer. He has published no refereed articles for work done since his departure from Cetus.

Chapter Five

1. Erlich was especially interested in the relation of HLA polymorphism to autoimmune maladies such as insulin-dependent diabetes. In order to learn more about the location and polymorphisms involved, it was necessary to make genomic libraries of diabetic siblings. Sibs were used because they shared one HLA chromosome. Cloning and sequencing would then be used to assign variations to different chromosomes. While producing tantalizing results, this mapping was a laborious task. By looking at the distribution of alleles in patient populations and in a control, one could identify which alleles were significantly more common among the patients; it was assumed that these alleles conferred increased risk. At Cetus, this approach was applied first to Insulin Dependent Diabetes Mellitus (IDDM), or type I diabetes, an autoimmune disease. Serological HLA typing had suggested that certain types were associated with increased risk but at the PCR level enabled a more precise subtype in genetic terms of the serologic types. "The Class II histocompatibility molecules are dimers composed of heavy alpha and light beta chains. They are encoded by genes in the HLA complex. Most of the sequence diversity of Class II histocompatibility molecules is localized within a segment of the beta chain"; King and Stansfield 1990, 148. The core information in this section is based on an interview by the author with Henry Erlich, 23 June 1993.

2. Daniell 1994, 421–23.

3. The first publication from Erlich's lab reporting the use of PCR on the HLA gene complex appeared in the 5 September 1986 issue of *Science*.

4. "Episomes may behave (1) as autonomous units replicating in the host independent of the bacterial chromosome, or (2) as integrated units attached to the bacterial chromosome and replicating with it"; King and Stansfield 1990, 106.

5. The first publication on the application of PCR to infectious agents was submitted to the *Journal of Virology* in November 1986 (appearing in May 1987). Kwok et al. 1987.

6. Shing Chang, who was White's associate director of Research, would be promoted and become head of Research.

7. Robert Fildes, interview by author, 10 December 1993.

8. David Gelfand, interview by author, 8 July 1993.

9. White et al. 1990; Bruns, White, and Taylor 1991.

10. *Wall Street Journal,* 11 January 1989.

11. In 1990 his curriculum vitae described the position as follows: "Reporting to the President of the Roche Diagnostic Group and the Vice President, Exploratory Research for Hoffmann-La Roche; responsible for the Roche-Cetus diagnostics program of $6 million per year (35 people) and setting up a Roche diagnostics R&D facility and staff in California (30 people). Responsible for design and buildout of a 22,000 ft^2 lab, 65 scientists, $9 million annual budget and overall management of Roche-Cetus program." The buildout was the new Roche Diagnostic Research (later Roche Molecular Systems) laboratory in Alameda into which the operations were moved and expanded once Hoffmann-La Roche acquired the PCR division at Cetus as well as the exclusive rights to PCR.

12. Fildes is mistaken about the dates. They should be 1982 as the start, 1984 for the clinic, 1992 for approval.

13. Ron Cape never responded to several written invitations to present his views.

14. *Wall Street Journal,* 17 August 1990.

15. *Wall Street Journal,* 23 July 1991.

Conclusion

1. Randall Saiki, interview by author, 12 November 1993.

2. Henry Erlich, interview by author, 9 November 1993.

3. Tom White, interview by author, 28 November 1993.

4. Weber 1946 , 152.

5. Weber 1958, 51.

6. Henry Erlich, personal communication, 8 June 1995.

7. Weber 1946, 143.

8. John Dewey, "Logic of Judgments of Practice," Dewey 1953, 441–42.

9. Tim Appenzeller, "Democratizing the DNA Sequence," *Science* 247 (2 March 1990): 1030. The article includes a quote from Mullis: "I was pretty lucky. The first time I tried the reaction it worked."

10. Kary Mullis, preface to Mullis, Ferre, and Gibbs 1994, x.

11. Lévi-Strauss 1966, 16. The French reads: *"Dans son sens ancien, le verbe bricoler s'applique an jeu de balle et de billard, à la chasse et à l'équitation, mais toujours pour évoquer un mouvement incident: celui de la balle qui rebondit, du chien qui divague, du cheval qui s'écarte de la ligne droite pour éviter un obstacle"; La Pensée Sauvage* (Paris: Plon, 1962), 26.

12. Michael Lynch and Kathleen Jordan are studying this question.

Bibliography

Arnheim, Norman. 1983. "Concerted Evolution of Multigene Families." In *Evolution of Genes and Proteins,* edited by M. Nei and R. K. Koehn, 38–62. Sunderland, MA: Sinauer Associates, Inc.

Bruns, T. D., T. J. White, and J. E. Taylor. 1991. "Fungal Molecular Systematics." *Annual Review of Ecology and Systematics* 22:525–64.

Chien, A., E. B. Edgar, and J. M. Treia. 1976. "Deoxyribonucleic acid polymerase from the extreme thermophile Thermus aquaticus." *Journal of Bacteriology* 127:1550–57.

Collins, Harry. 1975. "The Seven Sexes: A Study in the Sociology of a Phenomenon, or The Replication of Experiment in Physics." *Sociology* 9:205–24.

Conner, Brenda, A. Reyes, C. Morin, K. Itakura, R. Teplitz, and R. Wallace. 1983. "Detection of sickle cell beta-s globin allele by hybridization with synthetic oligonucleotides." *Proceedings of the National Academy of Sciences U.S.A.* 80 (January): 278–82.

Crick, Francis. 1988. *What Mad Pursuit: A Personal View of Scientific Discovery.* New York: Basic Books.

Daniell, Ellen. 1994. "PCR in the Marketplace." In *PCR: The Polymerase Chain Reaction,* edited by K. Mullis, F. Ferre, and R. Gibbs, 421–36. Basel: Birkhauser.

Dewey, John. [1917] 1953. *Essays in Experimental Logic.* New York: Dover Books.

Dickson, David. 1988. *The New Politics of Science.* 2d ed. Chicago: University of Chicago Press.

Dreyfus, H. L., and P. Rabinow. 1979. *Michel Foucault: Beyond Structuralism and Hermeneutics.* Chicago: University of Chicago Press.

Eisenberg, Rebecca S. 1987. "Proprietary Rights and the Norms of Science in Biotechnology Research." *Yale Law Journal* 97, no. 2 (December): 177–231.

Erlich, Henry, ed. 1989. *PCR Technology: Principles and Applications for DNA Amplification*. New York: Stockton Press.

Guyer, R. L., and D. E. Koshland Jr. 1989. "The Molecule of the Year." *Science*, 22 December, 1543.

Hall, Stephen S. 1987. *Invisible Frontiers: The Race to Synthesize a Human Gene*. Redmond, WA: Tempus Press.

Holtzman, Neil A. 1989. *Proceed with Caution: Predicting Genetic Risks in the Recombinant DNA Era*. Baltimore and London: Johns Hopkins University Press.

Jacob, François. 1988. *The Statue Within*. New York: Basic Books.

Jameson, Fredric. 1991. *Postmodernism, or the Cultural Logic of Late Capitalism*. Durham, NC: Duke University Press.

Kaleidin, A. S., A. G. Siyusarenko, and S. I. Gorodetski. 1980. "Isolation and properties of DNA polymerase from extremely thermophilic bacterium Thermus aquaticus YT1." *Biokhimiya* 45:644–51.

Kay, Lily E. 1993. *The Molecular Vision of Life: Caltech, the Rockefeller Foundation, and the Rise of the New Biology*. New York: Oxford University Press.

Keller, Evelyn Fox. 1985. *Reflections on Gender and Science*. New Haven: Yale University Press.

———. 1992. *Secrets of Life, Secrets of Death: Essays on Language, Gender and Science*. New York: Routledge.

Kenney, Martin. 1986. *Biotechnology: The University-Industry Complex*. New Haven: Yale University Press.

King, R. C., and W. D. Stansfield. 1990. *A Dictionary of Genetics*. 4th ed. New York: Oxford University Press.

Kloppenberg, Jack Ralph Jr. 1988. *First the Seed: The Political Economy of Plant Biotechnology, 1492–2000*. Cambridge: Cambridge University Press.

Kohler, Robert. 1976. "The Management of Science: Warren Weaver and the Rockefeller Foundation Program in Molecular Biology." *Minerva* 14:249–93.

Kornberg, Arthur. 1989. *For the Love of Enzymes: The Odyssey of a Biochemist*. Cambridge: Harvard University Press.

Krimsky, Sheldon. 1982. *Genetic Alchemy: The Social History of the Recombinant DNA Controversy*. Cambridge, MA: MIT Press.

Kwok, Shirley, D. H. Mack, K. B. Mullis, B. Poiesz, G. Elrich, D. Blair, A. Friedman-Kien, and J. J. Sninsky. 1987. "Identification of Human Immunodeficiency Virus Sequences by Using In Vitro Enzymatic Amplification and Oligomer Cleavage Detection." *Journal of Virology* 61 (5): 1690–94.

Lévi-Strauss, Claude. 1966. *The Savage Mind*. Chicago: University of Chicago Press.

Merton, Robert. 1973. "The Normative Structure of Science." In Merton, *The Sociology of Science: Theoretical and Empirical Investigations*. Chicago:

University of Chicago Press. Originally published as "Science and Technology in a Democratic Order," *Journal of Legal and Political Sociology* 1 (1942): 115–26.

Mukerji, Chandra. 1989. *A Fragile Power: Scientists and the State.* Princeton: Princeton University Press.

Mulkay, Michael. 1980. "Interpretation and the Use of Rules: The Case of the Norms of Science." In *Science and Social Structure: A Festschrift for Robert K. Merton,* Transactions of the New York Academy of Sciences, edited by Thomas Gieryn, 2d ser., no. 39, 111–25.

Mullis, Kary. 1990. "The Unusual Origin of the Polymerase Chain Reaction." *Scientific American,* April, 56–65.

Mullis, Kary, F. Faloona, S. Scharf, R. Saiki, G. Horn, and H. Erlich. 1986. "Specific Enzymatic Amplification of DNA in Vitro: The Polymerase Chain Reaction." *Cold Spring Harbor Symposium in Quantitative Biology* 51:263–73.

Nietzsche, Friedrich. [1885] 1968. *Beyond Good and Evil.* In *Basic Writings of Nietzsche,* edited by Walter Kaufmann. New York: The Modern Library.

Office of Technology Assessment (OTA). 1984a. *Commercial Biotechnology: An International Assessment.* Washington, DC: U.S. Government Printing Office.

———. 1984b. *Technology, Innovation, and Regional Economic Development.* Washington, DC: U.S. Government Printing Office.

———. 1988. *New Developments in Biotechnology: Ownership of Human Tissues and Cells.* Washington, DC: U.S. Government Printing Office.

Oste, Christian. 1989. "PCR Automations." In *PCR Technology: Principles and Applications for DNA Amplification,* edited by Henry Erlich, 23–30. New York: Stockton Press.

Panem, Sandra. 1984. *The Interferon Crusade.* Washington, DC: The Brookings Institute.

Pauly, Phillip J. 1987. *Controlling Life: Jacques Loeb and the Engineering Ideal in Biology.* New York: Oxford University Press.

Rabinow, Paul. [1989] 1995. *French Modern: Norms and Forms of the Social Environment.* Reprint. Chicago: University of Chicago Press.

———. 1992. "Artificiality and Enlightenment: From Sociobiology to Biosociality." In *Incorporations,* edited by J. Crary and S. Kwinter, 234–52. New York: Zone.

Rheinberger, Hans-Jorg. 1992. *Experiment, Differenz, Schrift, Zur Geschichte epistemischer Dinge.* Warburg: Basiliskenpresse.

Root-Bernstein, Robert Scott. 1989. *Discovering, Inventing and Solving Problems at the Frontiers of Scientific Knowledge.* Cambridge, MA: Harvard University Press.

Rosenberg, Steven A., and John Barry. 1992. *Transformed Cell: Unlocking the Mysteries of Cancer.* New York: Avon Books.

Saiki, R. K., S. Scharf, F. Faloona, K. Mullis, G. Horn, H. E. Erlich, and N. Arnheim. 1985. "Enzymatic Amplification of Beta-Globin Genomic Sequences and Restriction Site Analysis for Diagnosis of Sickle Cell Anemia." *Science* 230:1350–54.

Shapin, Steven. 1994. *A Social History of Truth: Civility and Science in Seventeenth Century England.* Chicago: University of Chicago Press.

Shilts, Randy. 1987. *And the Band Played On: Politics, People and the AIDS Epidemic.* New York: St. Martin's Press.

Smith, Jane S. 1990. *Patenting the Sun: Polio and the Salk Vaccine.* New York: Morrow.

Snow, C. P. [1959] 1964. *The Two Cultures and a Second Look.* Cambridge: Cambridge University Press.

Swann, John P. 1988. *Academic Scientists and the Pharmaceutical Industry.* Baltimore and London: Johns Hopkins University Press.

Teitelman, Robert. 1989. *Gene Dreams: Wall Street, Academia and the Rise of Biotechnology.* New York: Basic Books.

Thomas, Keith. 1983. *Man and the Natural World.* New York: Pantheon Books.

Watson, J. D., and F. H. Crick. 1953. "Molecular Structure of Nucleic Acids: A structure for Deoxyribose Nucleic Acid." *Nature* 171 (25 April): 737–38.

Weber, Max. 1946. "Science as a Vocation." In *From Max Weber: Essays in Sociology,* edited by H. Gerth and C. W. Mills. New York: Oxford University Press.

———. 1958. *The Protestant Ethic and the Spirit of Capitalism.* New York: Scribner's.

White, Tom, T. Bruns, S. Lee, and J. Taylor. 1990. "Amplification and direct DNA sequencing of fungal ribosomal RNA genes for phylogenetics." In *PCR Protocols: A Guide to Methods and Applications,* edited by H. Erlich, 315–22. New York: Stockton Press.

Wright, Susan. 1986a. "Recombinant DNA Technology and Its Social Transformation, 1972–1982." *Osiris,* 2d ser., no. 2, 303–60.

———. 1986b. "Molecular Biology or Molecular Politics? The Production of the Scientific Consensus on the Hazards of Recombinant DNA Technology." *Social Studies of Science* 16:593–620.

———. 1994. *Molecular Politics.* Chicago: University of Chicago Press.

Yoxen, Edward. 1982. "Giving Life a New Meaning: The Rise of the Molecular Biology Establishment." *Sociology of the Sciences* 6:123–43.

to clone IL-2, it should abandon IL-2 altogether and move on to
another protein. Other companies did decide to pursue that strategy; only Amgen and Cetus gambled and continued to pursue
their research and development. Although the two companies developed recombinant variants of IL-2 and were soon locked in a
legal battle with each other, Cetus had a lead over Amgen and
became the major supplier of IL-2 for clinical studies. Cetus's aggressive patent stance and its successful defense in the courts eventually forced Amgen out of the competition. Hoffmann-La Roche,
however, was a much stronger competitor than Amgen. At Cetus,
the internal debate continued over how much fighting and how
much negotiating to do with Roche.

Cetus had in fact approached Roche earlier about entering into
partnership, but Roche had declined; IL-2 was not a major product
for it and Cetus's mutein didn't interest it greatly. Cetus raised the
issue again in 1988, and this time Hoffmann-La Roche was indeed
interested, not in Cetus's IL-2 mutein, but in PCR. Having decided that PCR was an extremely promising property—the most
powerful DNA probe and amplification technology available—
Hoffmann-La Roche also decided that it wanted no three-way
partnership with Kodak or anyone else. Two agreements were
reached. Under the first, Roche would fund diagnostic research at
Cetus for five years at $6 million per year and pay a significant
royalty on the sale of jointly developed diagnostic products and
services. Roche also purchased warrants for a million shares of
Cetus stock at fifteen dollars per share, a figure three dollars above
the price on the stock market.[10] In the second agreement, Cetus
obtained freedom from suit under Roche's IL-2 patent, and the
two companies agreed to share clinical data.

As a deal between Cetus and Hoffmann-La Roche became increasingly likely, Hoffmann-La Roche asked Price for his recommendation of an in-house person who might be available to handle
the PCR operations if Hoffmann-La Roche acquired the rights.
Price suggested White. White was interviewed in January 1989
and hired in March 1989, one month after Cetus's contract with
Kodak expired.[11] White began his work, still housed in a Cetus
building, while waiting for the results of the patent trial between
Du Pont and Cetus, which would decide the ownership of and
commercial control over PCR.

ENDGAME

Increasingly during 1989 and the early months of 1990, there was a real question as to how long Cetus could continue as an independent company, given its continuing financial losses. There were many significant departures among the R&D staff, including those who were handling the IL-2 clinical trials. All the top R&D, clinical, and regulatory managers resigned. In White's view, these departures were directly the result of Fildes's management style, and it was difficult for him to envision a scenario in which new people could be hired to correct the situation. Ron Cape, who was at Cetus only infrequently, had shown that he was not going to take a stand against Fildes. Several former senior scientists at Cetus feel that Cape's passivity was certainly as responsible for the situation at Cetus during this period as Fildes's aggressive style and IL-2 strategy. The potential failure of Cetus posed a problem for Roche: if Cetus were to be acquired by another company, the PCR project could well be put in jeopardy.

During the early summer of 1990, yet another confrontation took place between Fildes and scientists from R&D. David Gelfand organized meetings with Cape and others to tell them that, in his view, Fildes was on a path that would destroy the company. Although they had ignored Gelfand's warnings three years earlier, it was more difficult to ignore them the second time around, after every top R&D manager in the company and other senior scientists had departed. Gelfand and the others were convinced that the company was not going to be rebuilt with Fildes at the helm. Gelfand bitterly remarks that at no point during this crucial period did the board consult Cetus's senior scientists.

INTERVIEW: JEFF PRICE

PAUL RABINOW   How was your relationship with Fildes?

JEFF PRICE   I worked with Bob about eight years; when I left Cetus, I'd been there about fourteen years, so I had pretty much understood what the opportunities and the limitations were of Bob's management of the company. I had considered